Audel™
Plumber's Pocket Manual

All New 10th Edition

Rex Miller
Mark Richard Miller
Joseph Almond, Sr.

Wiley Publishing, Inc.

Vice President and Executive Group Publisher: Richard Swadley
Vice President and Executive Publisher: Robert Ipsen
Vice President and Publisher: Joseph B. Wikert
Executive Editor: Carol A. Long
Editorial Manager: Kathryn A. Malm
Development Editor: Kevin Shafer
Production Editor: Vincent Kunkemueller
Text Design & Composition: TechBooks

Copyright © 2004 by Wiley Publishing, Inc. All rights reserved.

Published simultaneously in Canada

No part of this publication may be reproduced, stored in a retrieval system, or transmitted in any form or by any means, electronic, mechanical, photocopying, recording, scanning, or otherwise, except as permitted under Section 107 or 108 of the 1976 United States Copyright Act, without either the prior written permission of the Publisher, or authorization through payment of the appropriate per-copy fee to the Copyright Clearance Center, Inc., 222 Rosewood Drive, Danvers, MA 01923, (978) 750-8400, fax (978) 646-8700. Requests to the Publisher for permission should be addressed to the Legal Department, Wiley Publishing, Inc., 10475 Crosspoint Blvd., Indianapolis, IN 46256, (317) 572-3447, fax (317) 572-4447, E-mail: permcoordinator@wiley.com.

Limit of Liability/Disclaimer of Warranty: The publisher and the author make no representations or warranties with respect to the accuracy or completeness of the contents of this work and specifically disclaim all warranties, including without limitation warranties of fitness for a particular purpose. No warranty may be created or extended by sales or promotional materials. The advice and strategies contained herein may not be suitable for every situation. This work is sold with the understanding that the publisher is not engaged in rendering legal, accounting, or other professional services. If professional assistance is required, the services of a competent professional person should be sought. Neither the publisher nor the author shall be liable for damages arising herefrom. The fact that an organization or Web site is referred to in this work as a citation and/or a potential source of further information does not mean that the author or the publisher endorses the information the organization or Web site may provide or recommendations it may make. Further, readers should be aware that Internet Web sites listed in this work may have changed or disappeared between when this work was written and when it is read. For general information on our other products and services, please contact our Customer Care Department within the Unied States at (800) 762-2974, outside the United States at (317) 572-3993, or fax (317) 572-4002.

Trademarks: Wiley, the Wiley Publishing logo, Audel are trademarks or registered trademarks of John Wiley & Sons, Inc., and/or its affiliates. All other trademarks are the property of their respective owners. Wiley Publishing, Inc., is not associated with any product or vendor mentioned in this book.

Wiley also publishes its books in a variety of electronic formats. Some content that appears in print may not be available in electronic books.

Library of Congress Control Number:

ISBN: 0-7645-6995-3

10 9 8 7 6 5 4 3 2 1

Table of Contents

Acknowledgments	v
About the Author	vii
Introduction	ix

Part I Techniques, Installation, and Repair

1.	Tips for the Beginning Plumber	3
2.	Plumbing Safety and Tricks of the Trade	15
3.	Plumbing Tools	19
4.	Dental Office	33
5.	Working Drawings	45
6.	Roughing and Repair Information	57
7.	Outside Sewage Lift Station	105
8.	Pipes and Pipelines	119
9.	Vents, Drain Lines, and Septic Systems	129
10.	Lead Work	145
11.	Lead and Oakum Joints	157
12.	Silver Brazing and Soft Soldering	163

Part II Plumbing Systems

13.	Plastic Pipe and Fittings	183
14.	Cast-Iron Pipe and Fittings	211
15.	Copper Pipe and Fittings	279
16.	Water Heaters	311

17.	Water Coolers and Fountains	331
18.	Automatic Bathroom Systems	337

Part III General Reference Information

19.	Abbreviations, Definitions, and Symbols	375
20.	Formulas	385
21.	Metric Information Helpful to the Piping Industry	409
22.	Knots Commonly Used	425
23.	Typical Hoisting Signals	427
Index		429

Acknowledgments

As is the case with any book, many people have given generously of their time in assisting the authors. Their efforts, time, and suggestions have been of great value. We would like to take this opportunity to thank each of them for their contributions.

Manufacturers of equipment used in the plumbing trade are listed here and in the caption of each drawing, picture, or other type of illustration furnished. They make it possible to keep the book current and interesting. The listed manufacturers are but a sampling of all the people involved in producing the materials used in this trade. Without their assistance, this book could not have been authentic and would have neither appeal nor objectivity. To each and everyone, we thank you!

Rex Miller

Mark R. Miller

Joseph P. Almond, Sr.

American Standard Co.
Chase Brass and Copper Company
Copper Development Association
Crane Company
Dunham-Bush, Inc.
Electric Eel Manufacturing Co., Inc.
Eljer Plumbingware Company
General Engineering Company
Genova, Inc.
W.L. Gore and Associates
Josam Manufacturing Company
Jet, Inc.
Plumb Shop
Ridge Tool Company (Rigid)
Sloan Valve Company, M. Susan Kennedy
A.O. Smith Corporation

About the Authors

Rex Miller was a Professor of Industrial Technology at The State University of New York, College at Buffalo, for more than 35 years. He has taught on the technical school, high school, and college level for more than 40 years. He is the author or coauthor of more than 100 textbooks ranging from electronics to carpentry and sheet metal work. He has contributed more than 50 magazine articles over the years for technical publications. He is also the author of seven Civil War regimental histories.

Mark Richard Miller finished his B.S. degree in New York and moved on to Ball State University, where he obtained his master's and went to work in San Antonio. He taught in high school and went to graduate school in College Station, Texas, finishing the doctorate. He took a position at Texas A&M University in Kingsville, Texas, where he now teaches in the Industrial Technology Department as a Professor and Department Chairman. He has coauthored seven books and contributed many articles to technical magazines. His hobbies include refinishing a 1970 Plymouth Super Bird and a 1971 Roadrunner.

Joseph P. Almond, Sr., is a professional plumber in the trades and a long-time member of the United Association of Journeymen and Apprentices of the Plumbing and Pipe Fitting Industry of the United States and Canada.

Introduction

The Plumber's Pocket Manual was designed to assist in the training of apprentices in the plumbing and pipe-fitting trades. However, we have learned through the years that the do-it-yourselfer has made good use of the manual, and we'd like to extend a welcome to all who are willing and eager to keep the plumbing working in top-notch order everywhere. Throughout the years, this manual has also been a helpful companion on the job, making work easier and more accurate. A handy reference in the toolbox or in the back pocket is always a treasure when there is a demand for information in a hurry.

This book contains time-saving illustrations, plus many tips and shortcuts that lead to accurate installations. Such subjects as silver brazing, soft soldering, roughing and repair, lead work, and various pipe fittings and specifications beneficial to fabricators are included.

Also featured are illustrations on vents and venting, a dental office, outside sewage lift stations, the Sovent system, septic tanks, water heaters, the Sloan flush valve (both manual and automatic), solar system water heaters, Oasis water coolers, plumbing tools, lots of related math, metric information, and working drawings.

Whether used in the classroom or on the job, this book provides simplified and condensed information at your fingertips. The homeowner will also find some interesting topics and step-by-step procedures for almost every plumbing need around the house.

We enjoyed writing the book and trust that you will find it useful in whatever walk of life you pursue.

Part I

Techniques, Installation, and Repair

I. TIPS FOR THE BEGINNING PLUMBER

This chapter includes a variety of useful information. These facts that a plumber must know have been gleaned from years of plumbing experience. The apprentice (or even the journeyman) should find these tips useful.

Reading Blueprints

When a measurement is taken from a blueprint, it should be checked from both ends of the building to ensure accuracy.

Datum is an established level or elevation from which vertical measurements are taken. A *bench mark* (BM) is a measure on which all other elevations are based.

All buildings have a *base elevation* from which all other elevations and grades are determined. Some plans use 100.0 feet, while others use 0.00 feet.

For example, using 100.0 feet as the base level, a basement floor level of 91.5 feet would indicate a basement floor level that is 8 feet 6 inches below the first-floor level.

Bench marks permit the plumber to locate the elevations pertaining to the project at hand. A 96-foot or a 104-foot bench mark would indicate 4 feet below or 4 feet above the finish floor (FF). Examples of bench marks, thus, would include the following:

FF + 4.0 feet
FF − 2.0 feet

A *sectional elevation* drawing would provide the plumber with information as to the width and height of a specific portion of the structure.

Elevation measurements on piping plans are called *invert elevations*.

A figure in isometric position lies with one corner directly in front of you. The back corner is tilted to a 30° angle.

4 Chapter 1

A building plan may denote an invert elevation of 0.325 foot at one end of a pipeline and 0.400 foot at the opposite end—a difference of 0.75 foot. By multiplying 0.75 foot by 12, you will find that the difference in inches between the two points will be 9 inches (see Table 1-1).

Table 1-1 Converting Inches to Decimal Parts of a Foot

Inches	Parts of Foot	Inches	Parts of Foot
1	0.083	7	0.5833
2	0.1666	8	0.6667
3	0.25	9	0.75
4	0.333	10	0.8333
5	0.4167	11	0.9333
6	0.50	12	1.00

Shooting Grade Levels

Plumbers are often asked to set grade levels for various piping elevations, including catch basins, floor drains, and many other grade levels associated with their work. Therefore, it behooves every plumber and fitter to become familiar with this very important phase of the piping industry.

Two main parts to shooting grade levels are the dumpy level and the leveling rod.

The *dumpy level* (named after its inventor) is a surveyor's level with a short inverted telescope rigidly affixed. It rotates in a horizontal plane only. This level mounts onto a tripod.

The *leveling rod* is a graduated rod used in measuring the vertical distance between a point on the ground and the line of sight of a surveyor's level, or dumpy level. This rod is marked off in tenths and hundredths, and its scale is known as the *engineer's scale*.

Each foot on the leveling rod is divided into tenths and each tenth is divided into tenths (thus, 100 marks in all).

Every 10 marks have a number from 1 to 9; following each 9 will be the proper foot mark.

Always remember that each foot on the leveling rod equals 12 inches. Each tenth of this foot equals 1.2 inches. Each of the 100 marks contained in the engineer's foot equals 0.01 of a foot. The exception would be the leveling rods marked off in 0.02 of a foot. They have 5 marks to each tenth, 50 marks to each foot.

To determine inches from hundredths of a foot, you simply multiply by 12.

Fig. 1-1 shows a 1-foot section of a leveling rod marked off in 0.01 of a foot.

Note the arrow and where it is pointing (to 0.54 foot). Thus the calculation would be as follows:

$$\begin{array}{r} 0.54' \\ \times\ 12 \\ \hline 108 \\ 54 \\ \hline 6.48 \end{array} \text{ rounded to } 6\frac{1}{2}''$$

To convert decimal feet to inches, multiply by 12. You may then change decimal inches to inches and fractions.

Heating Systems

The compression tank plays an important part in the economical operation of a heating system. Heated water in the system expands. If no tank were installed, the expanding water would be forced out through the relief valve. In that case, cool water would be drawn in to replace the water lost by expansion.

Extra fuel is used to heat this cold water. Also, the constant adding of water brings in foreign matter (such as sediment or lime). This results in scaling of the boiler with an ever-increasing amount of fuel required for heating.

6 Chapter 1

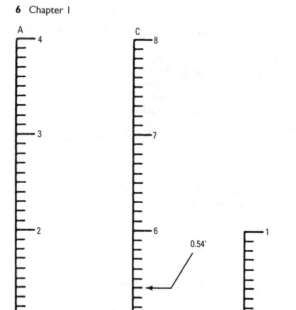

Fig. 1-1 Lower section of a leveling rod.

Note

Water in a heating system, when heated from 32°F to 212°F (0°C to 100°C) will expand approximately $1/23$ of its original volume.

Transfer of heat occurs in three ways:

- *Convection*—This is the transference of heat by the upward movement of a warm light air current.
- *Radiation*—This is the process in which energy in the form of heat is sent through space from atoms and molecules as they undergo internal change.
- *Conduction*—This is the transference of heat by the passage of energy, particle by particle.

Convection is the method used for transferring heat in a gravity domestic hot-water circulation system. Convection (or *circulating currents*) is produced because of the difference in weight of water at different temperatures.

Absolute pressure is *gage pressure* plus *atmospheric pressure*.

Water Heaters

The maximum acceptable temperature for domestic hot water is from 140°F to 160°F (60°C to 71°C). Use of automatic laundry and dishwashing machines makes 160°F (71°C) preferable. Temperatures above 160°F (71°C) are not recommended, because they cause increased corrosion, increased deposit of lime, waste of fuel, more rapid heat loss by radiation, scalding, and other accidents.

If a 30-gallon (113.56-liter) hot-water boiler is insulated with a tank jacket, 30 percent of the total amount of fuel usually burned can be saved.

If hot-water pipes are insulated, the heat loss from pipes is reduced by up to 80 percent.

There should be a minimum of 6 inches (15 cm) between an uninsulated water heater and any unprotected wood.

A special device called a *protector rod* is used to prevent corrosion in some water heaters.

Water Supply

Storage tanks up to 82-gallon (310.4-liter) capacity are tapped for 1-inch (25 mm) connections. Tanks over that size are generally tapped a minimum of 1¼ inches (32 mm).

The dip tube on a cold-water supply should terminate 8 inches (20 cm) above the bottom of the tank.

The standard length of asbestos cement water main pressure pipe is 13 feet or approximately 4 meters.

When water solidifies, it becomes lighter.

The amount of heat required to change ice to liquid water is 144 British thermal units (Btu) per pound (335 joules per kilogram).

Keep in mind that the installation of a water softener in a residential piping system causes some high-pressure loss.

Fixtures, Valves, and Fittings

When ordering piping elbows, an example would be as follows: 6 each, 1¼ inches (32 mm) copper, PVC, or galvanized 90° ells. If reducing elbows are ordered, list the largest measurement first: 6 each, 1¼ inches × 1 inch (32 × 25 mm) copper 90° reducing ells.

When ordering tees, you would begin by listing the largest measurement on the run or flow line—always listing the line measurement last. An example would be as follows: 6 each, 1¼ inches × 1 inch × 1½ inches (32 × 25 × 38 mm) copper tees (1½ inches is the branch line measurement).

In plumbing, the pipe size measurement given is always nominal pipe size (N.P.S.) inside diameter (ID). In air-conditioning and refrigeration, pipe and tubing are called and ordered by their outside diameter (OD) measurement.

Therefore, a ¼ inch (19 mm) copper pipe in plumbing would be called ⅞ inch (22 mm) pipe or tubing in refrigeration.

Flare fittings are sold and ordered by their OD measurement.

Brass fittings contain 85 percent copper, 5 percent zinc, 5 percent tin, and 5 percent lead.

Approximate heights above FF (or floor level) rims for plumbing fixtures are as follows:

- Sink—36 inches (91 cm)
- Built-in bathtub—16 inches (41 cm)
- Water closet—15 inches (38 cm)
- Lavatory—31 inches (79 cm)
- Wash or laundry tray—34 inches (86 cm)

An air gap of 1 inch (2.5 cm) to 2 inches (5 cm) between the flood level rim of a fixture and the water supply opening is considered safe.

A *vacuum* is a space entirely devoid of matter. A *partial vacuum* is a space where an air pressure exists that is less than atmospheric pressure. A vacuum relief valve should be installed on a copper tank to prevent collapse in the event of a vacuum occurrence. Suction pumps, barometers, and siphons depend on the natural pressure of the atmosphere in order to function.

A vacuum breaker should be at least 6 inches (15 cm) above the flood level rim, or 6 inches (15 cm) above the top of the unit.

Globe valves have a machined seat and a composition disc and usually shut off tight, while *gate valves* may leak slightly when closed (particularly if frequently operated) because of wear between the brass gates and the faces against which they operate. Globe valves create more flow resistance than gate valves.

When copper and steel contact each other, especially when dampness is present, a chemical action called *electrolysis* occurs.

A relief line from a relief valve is generally piped to the outside, 12 inches (305 mm) or less above ground level, elbow and nipple turned down. The nipple should not be threaded on outlet end.

The temperature-sensitive element of the relief valve should be installed directly in the tank proper so that it is in direct contact with the hot water.

The recommended and safest procedure is to place the relief valve in a separate tapping, either at the top of the tank or within 4 inches from the top if tapping is located at the side.

Common Terms
Fig. 1-2 shows some common terms to remember.

Scale Rules
Fig. 1-3 shows examples of scale rules.

Drains and Sewers
The *building drain* is the lowest horizontal piping inside the building. It connects with the building sewer.

The *building sewer* extends from the main sewer or other disposal terminal to the building drain at a distance of approximately 5 feet (152 cm) from the foundation wall.

Public sewer manholes can be used to verify main sewer elevations and direction of flow.

Gases found in sewer air are carbon monoxide, methane, hydrogen sulfide, carbon dioxide, gasoline, ammonia, sulfur dioxide, and illuminating gas.

Primary treatment in a sewage treatment plant removes floating and settleable solids. Secondary treatment removes dissolved solids.

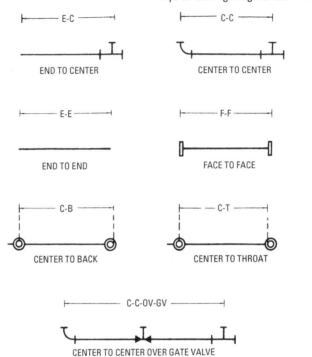

Fig. 1-2 Common terms to remember.

12 Chapter I

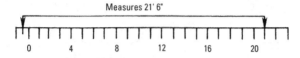

Scale: ⅛" = 1 Foot

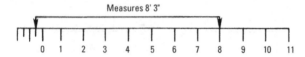

Scale: ¼" = 1 Foot

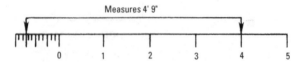

Scale: ½" = 1 Foot

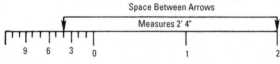

Scale: 1" = 1 Foot

Fig. 1-3 Scale rule examples for study.

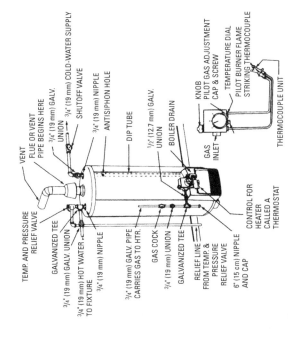

Fig. 1-4 Gas hot-water heater in average residence.

Gas Water Heaters

If there are no separate tappings on the water heater, then place the relief valve as shown in Fig. 1-4. However, use nipples as short as possible.

Never install a check valve in the water supply to a water heater, because it would confine pressure in the tank and result in an accident if the relief valve did not operate.

There is a small hole drilled in the dip tube near the top. This hole admits air to the cold-water piping to break the siphonic action.

The nipple and cap at the bottom of the tee where the gas supply turns into the heater form a dirt-and-drip pocket.

Following is the procedure for lighting a hot-water heater:

1. Turn gas cock on control to the "Off" position, and dial assembly to lowest temperature position.
2. Wait approximately 5 minutes to allow gas that may have accumulated in the burner compartment to escape.
3. Turn gas cock handle on control to the "Pilot" position.
4. Depress fully the set button and light the pilot burner.
5. Allow the pilot light to burn approximately 1 minute before releasing set button. If the pilot does not remain lighted, repeat the operation.
6. Turn the gas cock handle on the control to the "On" position and then turn the dial assembly to the desired position. The main burner will then ignite.

Note
Adjust the pilot burner air shutter (if provided) to obtain a soft blue flame.

2. PLUMBING SAFETY AND TRICKS OF THE TRADE

Plumbing is the art and science of creating and maintaining sanitary conditions in buildings used by humans. It is also the art and science of installing, repairing, and servicing in these same buildings a plumbing system that includes the pipes, fixtures, and appurtenances necessary for bringing in the water supply and removing liquid and water-carried waste.

The bathroom is not only the place where the "call of nature" is answered, but also the location utilized for privacy and showering (as well as dressing). A bathroom is a place where most of the plumbing is located in a house.

Bathrooms in large private homes were not unknown even in the eighteenth century. Beautifully equipped marble bathrooms are still preserved in several European palaces and mansions. However, it was not until the nineteenth century that bathrooms in private homes became commonplace. Fixtures generally included a toilet, bidet, washbasin, bath, mirror, and shelves.

In the twentieth century, the equipping of bathrooms became a separate industry with a wide variety of special forms of bathroom furniture and fixtures. The materials used included porcelain, enamel, plastic, wood, and stainless steel. People working with the installation and maintenance of bathrooms and kitchens are exposed to many dangers and safety hazards.

Applying good common sense and following a few safety rules can prevent a great deal of personal injury. This chapter discusses items that all plumbers should keep in mind.

Safety Tips

Electric power tools with abrasive cutting wheels should be properly grounded.

Wear eye protection and gloves.

Note
> People have been blinded by lime in the eyes, either as a powder or as mortar. When you get lime in your eyes, wash it out at once with flowing water. Check with an ophthalmologist (an eye doctor) to make sure the lime has not permanently caused eyesight damage. If lime touches your skin, wash with plenty of water. Then rinse the skin with vinegar. Coat the exposed skin with Vaseline mixed with baking soda (sodium bicarbonate).

When using a ladder, place the ladder base one-fourth of the ladder length away from the structure against which the ladder is leaning.

The maximum height for horses supporting scaffold platforms is 16 feet (4.88 m).

Vertical braces should be located every 8 feet (2.44 m) in trenches in hard, compact ground. Horizontal stringers in such soil conditions should be placed every 4 feet (1.22 m).

A good item to have on any job requiring the use of an open flame is a fire extinguisher. Open flames are used by plumbers in such tools as a Prestolite torch, a melting pot tank, and welding equipment. An approved carbon dioxide (CO_2) or a dry-powder fire extinguisher is usually useful. The dry-powder CO_2 extinguisher (which is actually composed of 99 percent baking soda and 1 percent drying agent) is also one of the best agents to use when fighting fires in or around electrical equipment.

Note
> A carbon tetrachloride (CCl_4) extinguisher at one time was considered the best agent to use, but it was found to release

Plumbing Safety and Tricks of the Trade 17

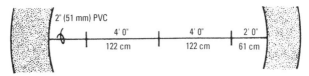

Fig. 2-1 Assembling 2-inch plastic pipe.

a deadly gas (phosgene). It can be especially dangerous if used in a confined place. Be careful because phosgene smells like newly mown hay. It was used as a poison gas in World War I.

Plumbing Tips

The maximum spacing for rainwater leaders (downspouts) is 75 feet (23 m). The recommended standard is 150 square feet (14 square meters) of roof area to 1 square inch (6.5 cm^2) of leader area.

The instrument for measuring relative humidity is called a hygrometer.

Bread can be packed in a water line to hold back water long enough to solder a joint where water continues to trickle, even if valves are shut off. The bread will dissolve when the water is turned on, and it can be flushed out of the system easily.

To replace a defective section or to add a fitting to a rigid pipeline, in most cases you will need to buckle the new portion back into the line in sections. For cast-iron pipe with lead joints and hard copper tubing, three sections and four joints will be necessary. On plastic pipe through 2 inches (51 mm), two sections 4 feet (122 cm) long and three joints will generally work. Fig. 2-1 shows an example of 2-inch (51 mm) plastic PVC pressure water line after it was repaired and reassembled.

3. PLUMBING TOOLS

A good plumber not only uses a variety of different tools but also knows a few inside tricks on getting the most from those tools. This chapter examines the basic tools of a plumber and presents a few miscellaneous tips on tools.

Basic Tools

The plumber's toolbox contains several different types of tools, including hammers, wrenches, pliers, tube cutters, augers, pipe reamers, and tube benders.

Fig. 3-1 shows a ballpeen hammer.

Fig. 3-1 Ballpeen hammer. *(Courtesy Ridge Tool Company)*

The pipe wrench shown in Fig. 3-2 allows one person to do the work of two. A short handle makes it easy to access frozen joints even in tight quarters.

Fig. 3-2 Pipe wrench—four sizes, 2 inches through 8 inches (51 through 203 mm). *(Courtesy Ridge Tool Company)*

The pipe wrench shown in Fig. 3-3 is known for the brutal punishment it can take. Before shipment, every wrench is

Fig. 3-3 Straight pipe wrenches—10 sizes, 6 inches through 60 inches (152 through 1524 mm). *(Courtesy Ridge Tool Company)*

work tested. The housing is replaced for free if it ever breaks or distorts. Replaceable jaws are made of hardened alloy steel. A full-floating hook jaw ensures instant grip and easy release. Spring suspension eliminates the chance that jaws could jam or lock on pipe. A handy pipe scale and large, easy-to-spin adjusting nut give fast one-hand setting to pipe size, and a comfort-grip malleable-iron I-beam handle has a convenient hang-up hole.

Figs. 3-4 through 3-18 show several other tools commonly found in a plumber's toolbox.

Fig. 3-4 Heavy-duty pipe cutter—cuts pipe $\frac{1}{2}$ inch through 2 inches (3 through 51 mm). *(Courtesy Ridge Tool Company)*

Plumbing Tools 21

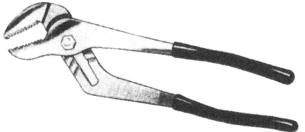

Fig. 3-5 Channel-lock pliers. *(Courtesy Ridge Tool Company)*

Fig. 3-6 Swaging tools. *(Courtesy Ridge Tool Company)*

22 Chapter 3

Fig. 3-7 Pipe and bolt threading machine for pipe ½ inch through 2 inches (3 through 51 mm). This machine cuts, threads, reams, and oils. *(Courtesy Ridge Tool Company)*

Fig. 3-8 Adjustable wrench (often called crescent wrench). *(Courtesy Ridge Tool Company)*

Plumbing Tools 23

Fig. 3-9 Internal wrench. This wrench holds closet spuds and bath, basin, and sink strainers through 2 inches (51 mm). Also handy for installing or extracting 1-inch through 2-inch (25-mm through 51-mm) nipples without damage to threads.
(Courtesy Ridge Tool Company)

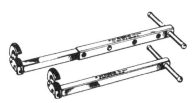

Fig. 3-10 Basin wrenches. The basin wrench has a solid 10-inch (254-mm) shank (shown on bottom of figure). Basin wrenches have telescopic shanks (top of figure) for four lengths from 10 inches through 17 inches (254 mm through 431 mm).
(Courtesy Ridge Tool Company)

24 Chapter 3

Fig. 3-11 Spud wrench—capacity 2⅝ inches (66.5 mm). *(Courtesy Ridge Tool Company)*

Fig. 3-12 Strap wrench—⅛ inch through 2 inches (3 mm through 51 mm). *(Courtesy Ridge Tool Company)*

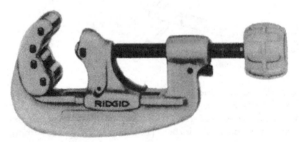

Fig. 3-13 Quick-acting tubing cutter—¼ inch through 2⅝ inches (6 mm through 66.5 mm). *(Courtesy Ridge Tool Company)*

Plumbing Tools 25

Fig. 3-14 Closet auger (used for water closet and urinal stoppage). *(Courtesy Ridge Tool Company)*

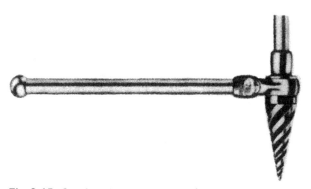

Fig. 3-15 Spiral ratchet pipe reamer—$1/2$ inch through 2 inches (3 mm through 51 mm). *(Courtesy Ridge Tool Company)*

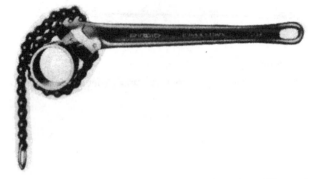

Fig. 3-16 Heavy-duty chain wrench—2 inches (51 mm). *(Courtesy Ridge Tool Company)*

Fig. 3-17 Torque wrench for cast-iron and no-hub soil pipe. Preset for 60 inch-lbs (67.9 N-m) of torque. *(Courtesy Ridge Tool Company)*

Plumbing Tools 27

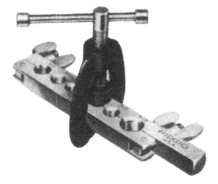

Fig. 3-18 Flaring tool—will flare tubing size from $3/_{16}$ inch (4.8 mm), $1/_4$ inch (6.4 mm), $5/_{16}$ inch (7.9 mm), $3/_8$ inch (9.5 mm), $7/_{16}$ inch (11.1 mm), $1/_2$ inch (12.7 mm), and $5/_8$ inch (15.9 mm). *(Courtesy Ridge Tool Company)*

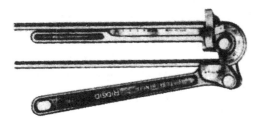

Fig. 3-19 Lever-type tube benders. *(Courtesy Ridge Tool Company)*

28 Chapter 3

Fig. 3-20 Straight snips. *(Courtesy Ridge Tool Company)*

Lever-type tube benders come in six sizes from $3/16$ inch through $1/2$ inch (5 mm through 13 mm) OD. They make fast, accurate bends on soft and hard copper, brass, aluminum, steel, and stainless steel tube. The form handle has *gain marks* for accurate tube measurements before cutting. Pieces can be precut to proper length, eliminating extra cutting and wasted material. The handles are wide apart when completing a full 180° bend (thus, no knuckle cracking). This machine cuts, threads, reams, and oils (see Fig. 3-19).

Figs. 3-20 through 3-23 show more plumbing tools.

Miscellaneous Tips on Tools

When a soldering iron is overheated, the normally bright areas will show a bluish tint, and the tinning on the bit will be dark, dull, and powdery in appearance.

The K-37 drain cleaner (see Fig. 3-24) speed-cleans $3/4$-inch (19-mm) through 3-inch (76-mm) lines without removing trap or crossbars. This rugged, compact unit with its dual-action clutch represents the latest in drain gun design. A slide-action handgrip permits operation on the fly while the

Plumbing Tools **29**

Fig. 3-21 Ratchet cutter—for cast-iron pipe 2 inches through 6 inches (51 mm through 152 mm). *(Courtesy Ridge Tool Company)*

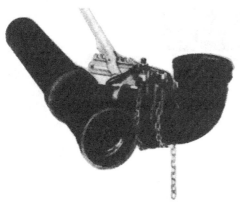

Fig. 3-22 Soil pipe assembly tool for pipe 2 inches through 8 inches (51 mm through 203 mm). *(Courtesy Ridge Tool Company)*

30 Chapter 3

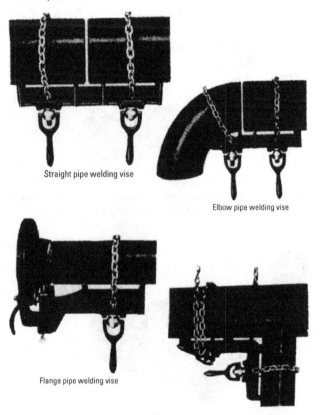

Straight pipe welding vise

Elbow pipe welding vise

Flange pipe welding vise

Angle pipe welding vise

Fig. 3-23 Pipe welding vises. *(Courtesy Ridge Tool Company)*

Plumbing Tools 31

Fig. 3-24 K-37 drain cleaner. *(Courtesy Ridge Tool Company)*

drum rotates. There is no need to stop the unit to advance or retract the cable. The knurled spin chuck is used for tough obstructions. A positive clutch lock transfers maximum torque to the stop-page and absorbs contact shock before it gets into the drum. Drum capacity is 35 feet (10.6 m) with $3/8$-inch (9.5-mm) cable, and 50 feet (15 m) with $5/16$-inch (8-mm) cable.

A friction clamp for use with brass pipe in a regular pipe vise is made by cutting a pipe coupling in half lengthwise and then lining it with sheet metal.

Use light machine oil when oiling a rule.

Hacksaw blade manufacturers recommend that a blade with 24 teeth per inch (10 teeth per cm) be used for cutting angle iron or pipe. To cut hanger rod, 18 teeth per inch (7 teeth per cm) is satisfactory. For light-gage band iron and thin-wall tubing, 32 teeth per inch (13 teeth per cm) is best.

The size of a pipe wrench is measured from the inside top of the movable jaw to the end of the handle with the wrench fully opened.

The wrench size for a flange bolt is *bolt size* $\times$ 2 + $\frac{1}{8}$ inch. In metric terms, it is *bolt size in millimeters* $\times$ 2 + 3 mm.

Acquire the habit of using two wrenches when tightening or loosening pipe to avoid many unnecessary problems.

4. DENTAL OFFICE

The dental office presents the plumber with unique challenges and a specific set of requirements. This chapter examines those challenges and requirements.

Dental Office Piping

This section provides guidelines for and examples of a typical dental chair installation.

Fig. 4-1 shows the DentalEZ Model CMU-D chair-mounted delivery system.

Fig. 4-2 shows a Hustler II compressor. It is one of the largest air power plants specifically designed for the dental profession. Its 30-gallon (113.6-liter) tank and powerful 12.2 standard cubic feet per minute (scfm) make it ideal for multiple clinics or group practices. This heavy-duty compressor operates with two $1\frac{1}{2}$-horsepower (hp) motors and pumps working simultaneously. The Hustler II is equipped with separate on-off switches for each compressor.

All Hustler models are equipped with an automatic tank drain that removes condensate at the end of each pumping cycle. Each Hustler is also equipped with a pump head unloader to provide easier starts and longer compressor life. A speed-reducing dual belt automatically maintains proper belt tension on each compressor.

The Hustler II is 45 inches (114 cm) wide, 32 inches (81 cm) high, and 19 inches (48 cm) deep.

Fig. 4-3 shows a deaquavator in connection with the Hustler II compressor. This deaquavator provides desert-dry air by removing more than 90 percent of all humidity and oil vapors from up to 10 cubic feet, or 283,168 cubic centimeters (cm^3), of air per minute. This very dry air minimizes the most frequent cause of costly hand-piece repair. The deaquavator operates on the same principle as a household refrigerator. It provides trouble-free performance for long periods.

34 Chapter 4

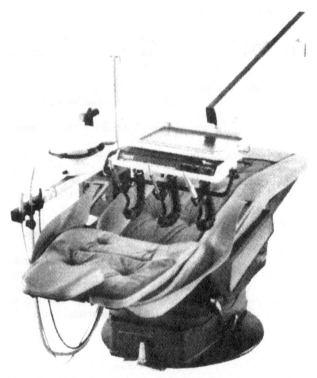

Fig. 4-1 DentalEZ chair. *(Courtesy DentalEZ)*

Fig. 4-2 Hustler II compressor. *(Courtesy DentalEZ)*

Fig. 4-3 Deaquavator. *(Courtesy DentalEZ)*

36 Chapter 4

The deaquavator is 18 inches (457 mm) wide, 14 inches (356 mm) high, and 13 inches (330 mm) deep. It weighs 62 pounds (28 kg). It operates on a 120-volt, 60-Hz, single-phase power source.

Fig. 4-4 provides a plan view of a dental chair layout, showing minimum location from walls and plumbing fixtures. Fig. 4-5 illustrates air piping from the compressor through the deaquavator to the dental chairs. Fig. 4-6

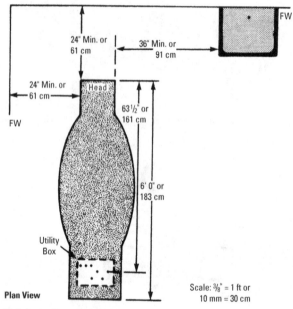

Fig. 4-4 Dental chair layout.

Dental Office **37**

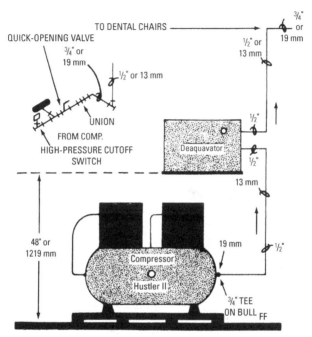

Fig. 4-5 Air piping to dental chairs.

38 Chapter 4

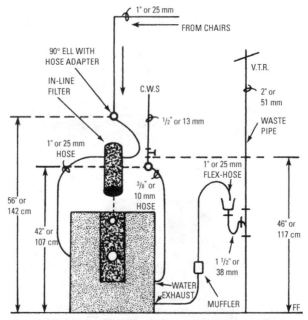

Fig. 4-6 Dynamic dual evacuation system.

Dental Office 39

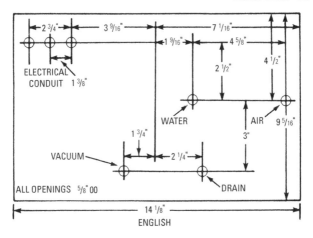

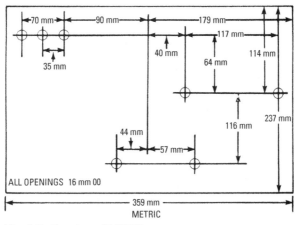

Fig. 4-7 Template of USC III.

40 Chapter 4

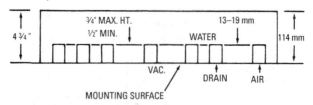

Fig. 4-8 Elevation view of utility layout.

illustrates the piping arrangement from the evacuator to the dental chairs. No part of the exhaust line shown in Fig. 4-6 should be more than 3 feet (91 cm) above the level of the waste connection on the vacuum pump. Also, there should be a 1-inch slope per 20 feet (25 mm every 6 meters) toward the vacuum producer. The evacuator can be located wherever it is convenient, or close to the compressor.

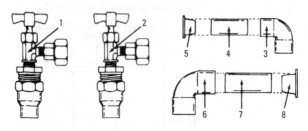

UTILITY STOP KIT NO. 3552-010

1. Water Stop
2. Air Stop
3. Copper Elbow Vacuum
4. Tubing Adapter for No. 3
5. Plug for Vacuum
6. Copper Elbow Gravity Drain
7. Tubing Adapter for No. 6
8. Plug for Gravity Drain

Fig. 4-9 Utility Stop Kit items.

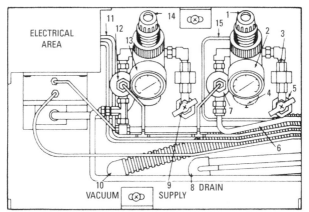

Fig. 4-10 Utility locations.

Fig. 4-7 shows the template used to position the supply piping to the dental chair utility box.

Fig. 4-8 shows an elevation view of the utility layout, and Fig. 4-9 displays the kit items. Fig. 4-10 shows a view of the utility locations. Fig. 4-11 illustrates the connections in the evacuation system shown in Fig. 4-6.

Dental Office Notes

Before proceeding with the plumbing or piping installation, consult all local applicable utility codes and regulations.

Central vacuum, gravity drain, air supply, and water supply piping through the floor should be $1/2$-inch-ID (13-mm-ID) rigid copper tubing.

On the air line from the compressor to the utility, rigid copper tubing (type L with brazed joints) is recommended. The air supply should be 80 to 100 psi, or 552 to 689 kPa.

42 Chapter 4

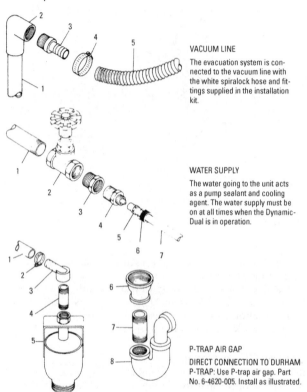

VACUUM LINE

The evacuation system is connected to the vacuum line with the white spiralock hose and fittings supplied in the installation kit.

WATER SUPPLY

The water going to the unit acts as a pump sealant and cooling agent. The water supply must be on at all times when the Dynamic-Dual is in operation.

P-TRAP AIR GAP

DIRECT CONNECTION TO DURHAM P-TRAP: Use P-trap air gap. Part No. 6-4620-005. Install as illustrated.

Fig. 4-11 Vacuum system connections.

On the drain from the floor, you should provide a 1½-inch or 2-inch (4-cm or 5-cm) P-trap below the floor, reducing the pipe as it goes through the floor.

The water supply pressure should be 40 to 45 psi, or 276 to 310 kPa.

5. WORKING DRAWINGS

This chapter examines working drawings a plumber might encounter when working with fresh-air systems, electric cellar drains, sand traps, trap seals, trailer connections, acid-diluting tanks, and residential garbage disposals.

Fresh-Air Systems

These systems are installed in places where food is sold, stored, handled, manufactured, or processed, such as restaurants, cafes, lunch stands, dairies, bakeries, and so on.

Note
A fresh-air master trap shall not be less than 4 inches (102 mm), as shown in Fig. 5-1.

Note
A fresh-air inlet may extend through a building wall approximately 12 inches (31 cm) above grade or extend through the roof. This inlet and auxiliary vent shall not be connected to any sanitary vent stack. Check local code.

Fig. 5-2 shows a typical grease trap installation. No sink trap is needed when the sink is connected to a grease trap. The interceptor is a trap itself and will prevent sewer gases from entering the house or building.

If a trap were added to this installation, it would restrict the flow and eventually cause a stoppage. Not only is this poor practice (*double trapping*, as it is called), but it is restricted in most plumbing codes.

A vent is provided on the outlet, or sewer side, to prevent siphonage of the contents of the grease trap.

Grease traps should be installed so that they provide access to the cover and a means for servicing and maintaining the trap.

46 Chapter 5

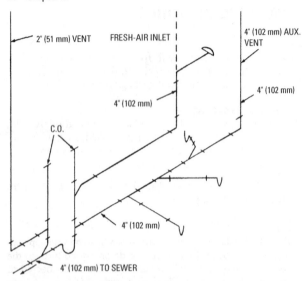

Fig. 5-1 Typical fresh-air system.

Check the local code in your area and the procedure set up by the administrative authority.

Electric Cellar Drains
Fig. 5-3 shows a working drawing for an electric cellar drain.

Sand Traps
Sand traps are generally used in filling stations, garages, and poultry houses—places where water carries sand, foam, refuse, or other material that would normally clog an ordinary drain. In the detail provided in a typical sand trap

Working Drawings 47

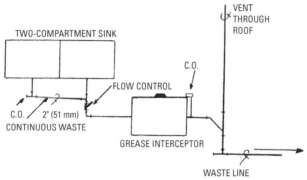

Fig. 5-2 Grease trap installation.

installation shown in Fig. 5-4, 33½ inches (85 cm) would be the distance from the bottom of the pit to the bottom of the inlet or invert.

When the sand trap is located in an open area such as a wash rack, slab, or similar place, the 2-inch (51-mm) vent may be omitted. Check the local code.

The inside dimension shown in Fig. 5-4 is the minimum size in many areas: 24 inches × 24 inches (61 × 61 cm).

Trap Seals

As shown in Fig. 5-5, the P-trap seal is measured from the top dip to the crown weir.

To protect a trap water seal from evaporation in a building that will be unoccupied for a period of time, pour a thin film of oil into the trap. During cold seasons, in unheated buildings, water should be drained and replaced with kerosene.

Trap seals may be lost by siphonage, evaporation, capillary attraction, or wind blowing.

48 Chapter 5

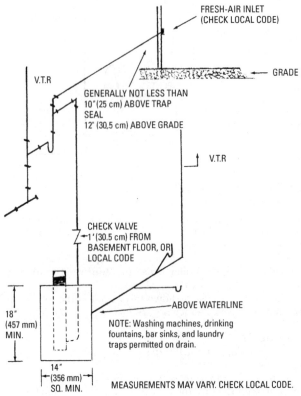

Fig. 5-3 Installation drawing for an electric cellar drain.

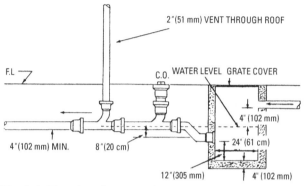

Fig. 5-4 Typical sand trap installation.

The standard P-trap seal is 2 inches (51 mm). Seals over 2½ inches (64 mm) are called *deep seals*.

Trailer Connections

As shown in the working drawing of a rough-in for a mobile home or trailer in Fig. 5-6, the P-trap is at least 18 inches (46 cm) below grade, and the inlet not more than 4 inches (102 mm) above grade. The connection from the trailer to the inlet should not exceed 8 inches (244 cm). The minimum distance between sewer and water connection should be 5 inches (152 cm). Check the local code.

Acid-Diluting Tanks

Fig. 5-7 shows a working drawing for an example of an acid-diluting tank installation.

Residential Garbage Disposals

The bottom half of Fig. 5-8 shows three methods of roughing-in the waste for a two-compartment sink with a disposal. The

50 Chapter 5

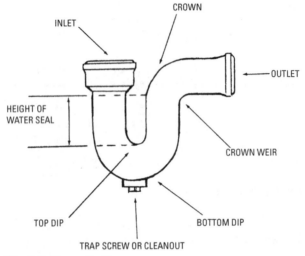

Fig. 5-5 Parts of a P-trap.

illustration on the left is common in alteration work, showing a double drainage wye inserted into the waste line so that another waste arm and separate trap to provide for the waste of the disposal can be run. In the middle illustration, the lower sanitary tee is roughed-in at 16 inches (41 cm) to 18 inches (46 cm) above the finished floor.

Check the local code for the proper sizing of the waste pipes, traps, and vent pipes.

Fig. 5-9 shows the parts used to mount a disposal. After the lock ring has been inserted in place, you will then begin tightening the screws in the flange. Do not tighten them completely until the disposal is set up and positioned for the waste trap.

Working Drawings **51**

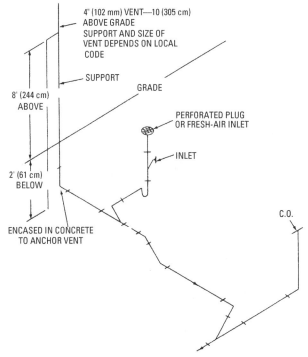

Fig. 5-6 Drawing of rough-in for mobile home or trailer.

52 Chapter 5

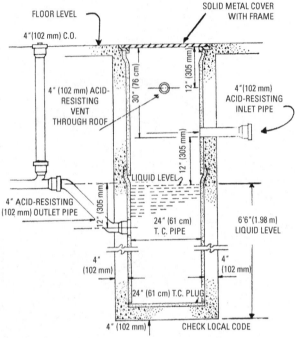

Fig. 5-7 Cross-sectional drawing of a typical example of an acid-diluting tank installation.

Working Drawings 53

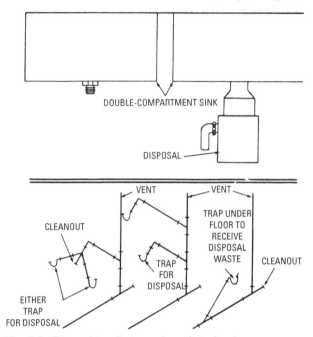

Fig. 5-8 Disposal installation and rough-in drawings.

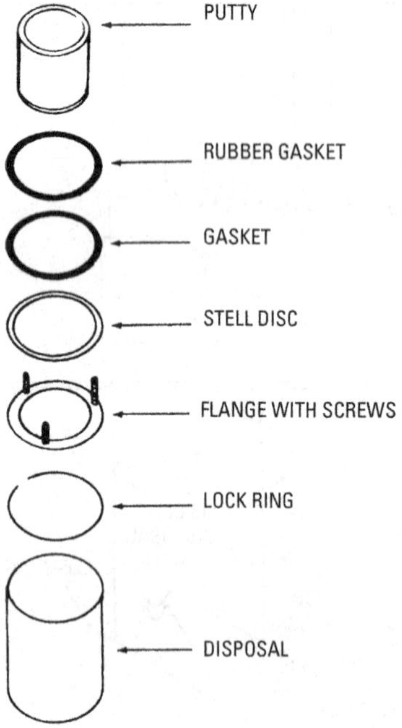

54 Chapter 5

- PUTTY
- RUBBER GASKET
- GASKET
- STELL DISC
- FLANGE WITH SCREWS
- LOCK RING
- DISPOSAL

Fig. 5-9 Exploded view of residential disposal mounting.

Once the disposal is mounted and positioned, you can then proceed to secure it to the waste line. If soldering must be done, be sure to tighten all the slipnuts first and do the sweating or soldering last.

On two-compartment sinks, the disposal is roughed low, around 16 inches (41 cm), the sink itself around 20 inches (50 cm) or 21 inches (53 cm).

No garbage disposal unit shall be installed on any indirect fresh-air waste system or into any grease interceptor.

Cold water must be used with a disposal, because it congeals grease particles, mixing them with food particles being flushed down the drain.

Hot water liquefies grease, and if constantly used, a stoppage would eventually occur because of accumulated coatings of grease.

An exception to this occurs when a drainage line has just been unstopped by a drain cleaner (especially a drain line from a kitchen sink). In this case, it is always good practice to flush the newly opened drain with hot water for at least 5 minutes. In this time, all the loosened-up grease that had accumulated inside the drainpipe will be liquefied by the hot water and immediately flushed down the drain.

6. ROUGHING AND REPAIR INFORMATION

This chapter provides valuable information about roughing in and repairing common fixtures such as bathtubs, sinks, water closets, urinals, faucets, washing machines, and whirlpool baths.

Bathrooms

This section provides a general guide for familiarization and for use where rough-in sheets pertaining to common bathroom fixtures are not available.

Fig. 6-1 shows the general location of fixtures in a bathroom. Typical connections to fixtures are illustrated in Fig. 6-2, and the schematic drawing in Fig. 6-3 shows the sewer and vent layout and connections, using cast-iron pipe and fittings with hubs. Of course, local codes should always be checked.

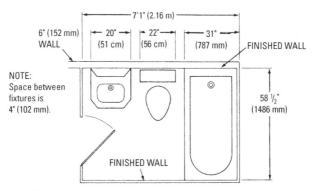

Fig. 6-1 Typical bathroom fixture locations.

58 Chapter 6

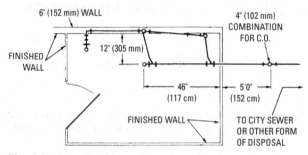

Fig. 6-2 Typical bathroom fixture connections.

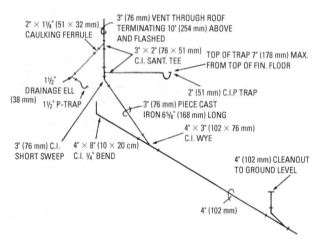

Fig. 6-3 Schematic drawing of a typical bathroom sewer and vent system.

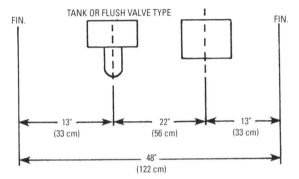

Fig. 6-4 Lavatory and water closet rough-in dimensions.

A bathroom that includes one water closet and one 20-inch (51-cm) lavatory can be placed in a minimum space of 48 inches (122 cm), finish to finish (Fig. 6-4).

For lavatories (Fig. 6-5 and Fig. 6-6), hot and cold water will rough at $20\frac{1}{2}$ inches (521 mm) if using speedy supplies. If brass nipples are used, check your roughing-in sheets. For the handicapped, the flood rim should be at 36 inches.

The rough-in for waste is $17\frac{1}{2}$ inches (444 mm) for a pop-up waste or drain plug. The backing centerline (CL) for the mounting bracket is 31 inches (787 mm), using a 2-inch × 10-inch (5-cm × 25-cm) support.

The waste location for corner lavatories is $17\frac{1}{2}$ inches (444 mm) from finished floor, $6\frac{3}{4}$ inches (171 mm) from corner to center (left or right). Using speedy supplies, the water location is $20\frac{1}{2}$ inches (521 mm) from the finished floor, 7 inches (178 mm) from corner to center, with hot water on the left and cold water on the right. The CL of the backing is 32 inches (813 mm), using a 2-inch × 8-inch (5-cm × 20-cm) support.

60 Chapter 6

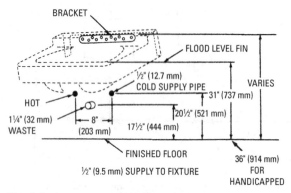

Fig. 6-5 Lavatory rough-in dimensions.

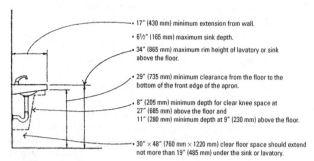

Fig. 6-6 Specifications for lavatory and sink installation.

For sanitary lavatories, the waste location is not more than 16 inches (406 mm) from finished floor, and the water location not more than 18 inches (457 mm) from finished floor.

To install a pop-up drain, follow these steps (see Fig. 6-7):

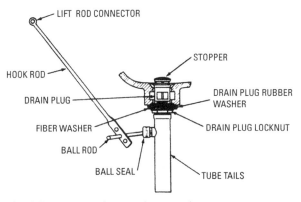

Fig. 6-7 Drawing of a typical pop-up drain.

1. Remove the drain plug from the tube tail. Detach the locknut, rubber washer, and fiber washer.
2. Insert the drain plug through the drain hole of the lavatory. Use plumber's putty underneath the flange of the plug. Attach the rubber washer, fiber, or metal washer and locknut.
3. Assemble the tube tail to the drain plug. Use a good pipe joint compound, and tighten. Turn the drain so that the side hole in the tube is pointed to the rear of the lavatory. Tighten the locknut.
4. Assemble the ball rod assembly to the hole in the side of the tube. Tighten loosely by hand.

62 Chapter 6

5. Insert the stopper into the drain.
6. Attach the hook rod as shown in Fig. 6-7 and tighten the setscrew so that the drain works properly by operating the knob.

Kitchen Sinks

The waste line should be $1\frac{1}{2}$ inches (38 mm) Standard Plumbing Code (S.P.C.) and located $22\frac{1}{4}$ inches (565 mm) from the finished floor, 8 inches (203 mm) off the centerline of a double-compartment sink. A single-compartment sink should be roughed in at $25\frac{1}{4}$ inches (641 mm). Hot- and cold-water lines should be 23 inches (584 mm) from the finished floor—hot 4 inches (102 mm) to the left of the centerline, and cold 4 inches (102 mm) to the right.

If one compartment of a two-compartment sink is to be provided with a garbage disposal, the waste line should rough in at 16 inches (406 mm) above the finished floor.

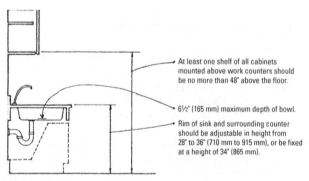

Fig. 6-8 Americans with Disabilities Act (ADA) guidelines for sink installation.

Fig. 6-6 shows the installation specifications for the kitchen sink. The Americans with Disabilities Act specifies a slightly different height than normal so that those seated in a wheelchair can reach the sink. This is shown in Fig. 6-8.

Service Sinks

The waste line or trap standard from these sinks is generally 3 inches (76 mm) S.P.S., and it roughs in at 10½ inches (267 mm) above the finished floor. The water lines are generally roughed in at 6 inches (152 mm) from the finished floor—hot 4 inches (102 mm) to the left, cold 4 inches (102 mm) to the right. Fig. 6-9 illustrates the installation of a sink with a U-trap. The faucet on a service sink may be sturdier and have a hook on the end to hold the bucket handle. In most instances, the service sink is mounted within a support framework (either a cabinet or steel support structure).

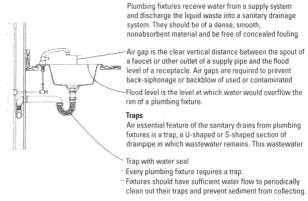

Plumbing fixtures receive water from a supply system and discharge the liquid waste into a sanitary drainage system. They should be of a dense, smooth, nonabsorbent material and be free of concealed fouling

Air gap is the clear vertical distance between the spout of a faucet or other outlet of a supply pipe and the flood level of a receptacle. Air gaps are required to prevent back-siphonage or backblow of used or contaminated

Flood level is the level at which water would overflow the rim of a plumbing fixture.

Traps

An essential feature of the sanitary drains from plumbing fixtures is a trap, a U-shaped or S-shaped section of drainpipe in which wastewater remains. This wastewater

Trap with water seal
· Every plumbing fixture requires a trap.
· Fixtures should have sufficient water flow to periodically clean out their traps and prevent sediment from collecting.

Fig. 6-9 Cross-sectional view of a mounted sink with a U-trap.

Water Closets

Water closet bowls (floor-mounted) are generally roughed in at 10 inches or 12 inches (254 mm or 305 mm) from the finished wall on centerline, using a 3-inch or 4-inch (76-mm or 102-mm) S.P.S. waste pipe. Wall-hung water closet bowls will generally rough in at $4\frac{1}{2}$ inches to $5\frac{1}{2}$ inches (114 mm to 140 mm) above the finished floor.

Water closets are supplied with water for flushing by either a flush tank as shown in Fig. 6-10 or a flush valve.

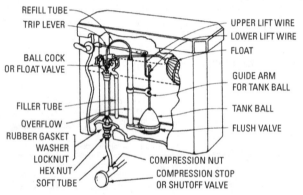

Fig. 6-10 Drawing of a water closet flush tank.

A water closet flush valve supply pipe is located $4\frac{3}{4}$ inches (121 mm) to the right from the centerline. This cold-water supply pipe is 1 inch (25 mm) S.P.S., 26 inches (660 mm) from the finished floor, or $20\frac{1}{2}$ inches (521 mm) from the center of a wall-hung water closet waste line to the center of the water supply line. In hospitals and nursing homes, the cold-water supply pipe should be roughed in at 36 inches (914 mm) above the finished floor.

When flush water is supplied by a flush tank, the water supply piping is usually roughed in 6 inches (152 mm) off the centerline to the left when facing the bowl, and 6 inches (152 mm) from the finished floor when using speed supplies or soft tubing. The size of the supply pipe should be $1/2$ inch (13 mm) ID.

When installing a water closet (floor-mounted) to a closet floor flange, whether plastic, brass, or cast iron, carefully tighten the closet nuts. Tighten both sides evenly, and tighten just enough so that the bowl does not rock. As soon as the nuts are drawn up snug, sit down on the bowl to settle the wax seal into place and then snug the nuts up a little more. Remember, drawing the nuts up too tightly will crack the bowl!

On the water closet flush valves and lavatory supply tubes, tighten from the top down. This will avoid possible leaks and a waste of time.

To replace a ball cock or float valve in a water closet flush tank, follow these steps (see Fig. 6-10):

1. Close the valve that supplies water to the tank.
2. Flush the tank and remove the remaining water with a sponge or rag.
3. Holding the float valve with one hand to prevent its turning, begin loosening the hex nut or the nut securing the supply tube to the float valve.
4. Begin loosening the locknut and lift the float valve out of the tank.
5. Before placing a new float valve in the tank, be sure the spot is clean and free of dirt or rust.
6. To reassemble, follow instructions in reverse order. Be sure to use pipe dope on threads.

7. When a new float valve is installed, take the refill tube, screw it into the opening provided, hold the refill tube at one end, and then bend the other end until it enters the overflow pipe.

You may have occasion to replace a worn or broken flush lever handle that flushes the tank. The first thing you would do is close the valve that supplies the water to the tank. Next, unhook the wire or chain from the old lever. Then, remove the flush handle nut inside of the tank (new and old nuts are threaded left or counterclockwise). You can now remove the old handle and lever. Put the new lever in the opening, slide the nut up the lever (large round side first), and bolt the handle in place with the lever in the horizontal position. Last, attach the tank ball wire. In many modern tanks, this is the chain to the Corky, which replaces the tank ball and the lift wires.

Bathtubs and Shower Stalls

To rough in a bathtub (see Fig. 6-11), the waste location is $1\frac{1}{2}$ inches (38 mm) off the rough wall on the centerline of the waste. The P-trap top should be no higher than 7 inches (18 cm) below floor level.

Note
Put the drain piece on last. The shower rod location is 76 inches (193 cm) high and 27 inches (69 cm) off the finished wall or against the outside tub rim.

Lead pans for shower stalls installed on new concrete floors should be given a heavy coating of asphaltum, both inside and out. The asphaltum protects the lead from corrosion during the curing period of the concrete caused by a chemical reaction that occurs between concrete, lead, and the water seeping through and contacting both.

Roughing and Repair Information **67**

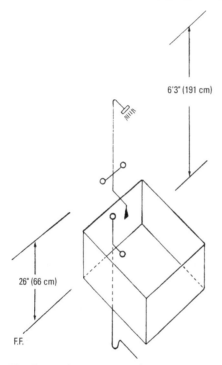

Fig. 6-11 Tub view in rough.

The installation of a typical tub trip waste and overflow is handled as follows. The numbers in parentheses refer to Fig. 6-12.

1. Remove the flathead screw (14) and perforated strainer plate (13) from the drain spud (12). Apply a small

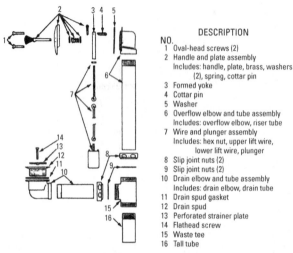

NO.	DESCRIPTION
1	Oval-head screws (2)
2	Handle and plate assembly Includes: handle, plate, brass, washers (2), spring, cottar pin
3	Formed yoke
4	Cottar pin
5	Washer
6	Overflow elbow and tube assembly Includes: overflow elbow, riser tube
7	Wire and plunger assembly Includes: hex nut, upper lift wire, lower lift wire, plunger
8	Slip joint nuts (2)
9	Slip joint nuts (2)
10	Drain elbow and tube assembly Includes: drain elbow, drain tube
11	Drain spud gasket
12	Drain spud
13	Perforated strainer plate
14	Flathead screw
15	Waste tee
16	Tall tube

Fig. 6-12 Exploded view of a typical tub trip waste overflow.

amount of putty to the underside of the drain spud.

2. Insert the drain spud (12) through the tub drain outlet from the inside. Place the drain spud gasket (11) on the face of the drain elbow as shown in Fig. 6-12. Proceed to tighten the drain spud (12) until it is secure. When the spud (12) is secure, the drain tube should point directly to the end of the tub.

3. Replace the perforated strainer plate and flathead screw in the drain spud.

4. Place the slip-joint nut and one slip-joint washer on the drain tube or shoe (10), and one slip-joint nut and one slip-joint washer on the riser tube (6).

5. Place the riser tube into the long end of the waste tee (15) and hand-tighten the slip-joint nut.
6. Place the washer (5) on the face of the overflow elbow and push the complete assembly onto the drain tube and hand-tighten the slip-joint nut.
7. Line up the washer with the overflow opening in the tub. Insert the plunger and wire inside the tub, and feed the plunger and wire through the opening until the handle and plate line up with the overflow opening in the tub.
8. Secure the plate in place by screwing two oval-head screws through the plate into the overflow elbow.
9. Wrench-tighten the two slip-joint nuts so that the drain tube and the riser tube are sealed to the waste tee.
10. The tub may now be placed in position with the tail tube or tailpiece (16) slipped into the drainage line or connected to it and sealed tight.

Depending on the size of the tub, occasionally the drain tube (10) and overflow tube (8) will need to be cut shorter.

Note
Wire and plunger assembly may need to be adjusted so that drain will work properly.

Typical Stall Urinals

The opening for a urinal should be approximately 24 inches (61 cm) wide and $18\frac{1}{2}$ inches (47 cm) from an abutting finished wall.

The 2-inch (51-mm) waste line should be $\frac{1}{4}$ inch (7 mm) to $\frac{3}{8}$ inch (10 mm) below the top line of the spud or strainer.

The top of the lip should be $\frac{1}{4}$ inch (7 mm) below the finished floor. Some codes call for the top of the lip to be above the finished floor (see Fig. 6-13). Check the local code.

70 Chapter 6

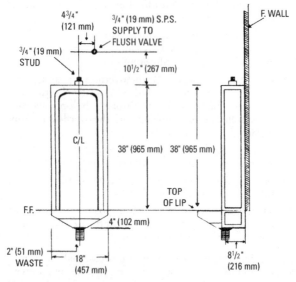

Fig. 6-13 Front and side views of a typical stall urinal.

Sharp sand should be packed under the urinal, and when the urinal is set, sand should be packed at least 1 inch (25 mm) up on the base of the urinal. When urinals are placed in a battery, spreaders are available; 3-inch (76-mm) spreaders are popular.

Sloan Flush Valves

All Sloan flush valves manufactured since 1906 can be repaired. It is recommended that when service is required, all inside parts be replaced. This restores the flush valve to like-new condition. To do so, order parts in kit form from your local dealer.

Figs. 6-14 through 6-17 show several views of the Sloan valve. This flush valve is in the process of being replaced by a completely automatic valve. The Optima Flushometer is discussed in Chapter 23. It is a good example of the automatic flush valve.

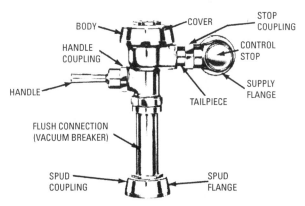

Fig. 6-14 Sloan flush valve.

Repairing Water Faucets and Valves

To repair water faucets and valves, follow these steps. The numbers in parentheses refer to Fig. 6-18.

1. Shut off the water supply valves or stops.
2. Remove the handle screw.
3. Remove the handle (2). Be sure the faucet is *not* completely closed before attempting to loosen the locknut

72 Chapter 6

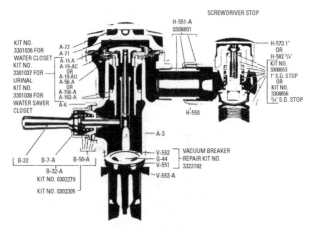

Fig. 6-15 Sloan (diaphragm-type) flush valve.

(3) (or, on many faucets, a packing nut) that will allow you to remove the stem (5). Open the faucet a quarter-turn and continue to check as the locknut is loosened. A crescent wrench should be used.

4. The stem (5) can then be removed.
5. Replace the washer at the bottom of the stem.

Note
If the Bibb screw holding the washer appears old and is difficult to remove, cut the washer out with a penknife. A pair of pliers can then be used, whereas a screwdriver may have turned off part of a screw head.

6. Before replacing the stem, examine the seat (10) located at the bottom (inside the valve body) where the washer seats. If the seat (10) appears rough, or a notch

Roughing and Repair Information 73

The repair kits and parts listed are designed to service all Sloan [diaphragm type] exposed and concealed flush valves. Each item has been identified by a part number along with a corresponding code number. To expedite your replacement requirements, order by code number.

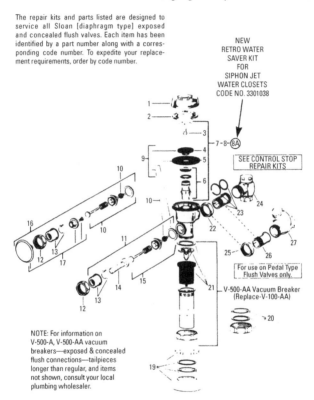

Fig. 6-16 Sloan (diaphragm-type) flush valve produced since mid-year 1971 (with parts list). *(Courtesy Sloan)* *(continued)*

1. 0301172 *A-72 CP Cover
2. 0301168 A-71 Inside Cover
3. 3301058 A-19-AC Relief Valve (Closet)—12 per pkg.
   3301059 A-19-AU Relief Valve (Urinal)—12 per pkg.
4. 3301111 A-15-A Disc—12 per pkg.
   0301112 A-15-A Disc (Hot Water)
5. 3301188 A-156-A Diaphragm w/A-29—12 per pkg.
   0301190 A-156-A Diaphragm (Hot Water)
6. 3301236 A-163-A Guide—12 per pkg.
7. 3301036 Inside Parts Kit for Closets, Service Sinks, Blowout, and Siphon Jet Urinals
8. 3301037 Inside Parts Kit for Washdown Urinals
8A. 3301038 Retro Water Saver Kit—delivers 3½ gal.
9. 3301189 A-156-AA Closet/Urinal Washer Set—6 per pkg.
10. 3302297 B-39 Seal—12 per pkg.
11. 3302279 B-32-A CP Handle Assem.—6 per pkg.
12. 0301082 *A-6 CP Handle Coupling
13. 0302109 B-7-A CP Socket
14. 0302274 B-32 CP Grip—12 per pkg.
15. 3302305 B-50-A Handle Repair Kit—6 per pkg.
16. 0303351 C-42-A 3" CP Push-Button Assem.
17. 3303347 3" CP Push-Button Replacement Kit
18. 3303396 C-64-A 3" Push-Button Repair Kit
19. 0306125 F-5-A ¾" CP Spud Coupling Assem.
    0306132 F-5-A 1" CP Spud Coupling Assem.
    0306140 F-5-A 1¼" CP Spud Coupling Assem.
    0306146 F-5-A 1½" CP Spud Coupling Assem.
20. 0306052 *F-2-A ¾" CP Outlet Coupling Assem.
    0306077 F-2-A 1" CP Outlet Coupling Assem.
    0306092 *F-2-A 1½" CP Outlet Coupling Assem. w/S-30
    0306060 *F-2-A 1¼" CP Outlet Coupling Assem.
    0306093 *F-2-A 1½" CP Outlet Coupling Assem.
21. 3323192 V-500-A & V-500-AA Vacuum Breaker Repair Kit
22. 0308676 *H-550 CP Stop Coupling
23. 0308801 *H-551-A CP Adj. Tail 2¹⁄₁₆" Long
24. 0308757 H-600-A 1" SD Bak-Chek CP Control Stop
    0308724 H-600-A ¾" SD Bak-Chek CP Control Stop
    0308881 *H-600-A 1" WH Bak-Chek CP Control Stop
25. 0308889 *H-600-A ¾" WH Bak-Chek CP Control Stop
26. 0308063 *H-6 CP Stop Coupling
27. 0308884 H-650-AG 1"SD Bak-Chek CP Control Stop
    0306882 *H-650-AG 1" WH Bak-Chek CP Control Stop

*Items also available in rough brass. Consult local plumbing wholesaler for code number.

Fig. 6-16 (*continued*)

Roughing and Repair Information **75**

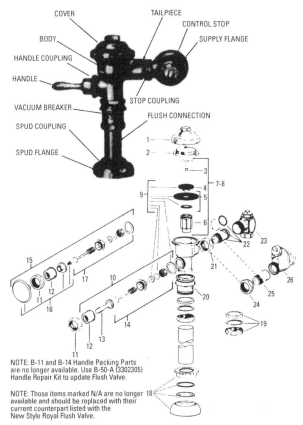

NOTE: B-11 and B-14 Handle Packing Parts are no longer available. Use B-50-A (3302305) Handle Repair Kit to update Flush Valve.

NOTE: Those items marked N/A are no longer available and should be replaced with their current counterpart listed with the New Style Royal Flush Valve.

Fig. 6-17 Sloan (diaphragm-type) flush valve produced prior to mid-year 1971 (with parts list). *(Courtesy Sloan)* (*continued*)

76 Chapter 6

1. Cover CP N/A; use 0301172 and 0301168.
2. Inside Brass Cover N/A; use 0301168 and 0301172.
3. A-19-A Brass Relief Valve N/A; use 3301058 or 3301059.
4. 3301111 A-15-A Disc—12 per pkg.
   0301112 A-15-A Disc (Hot Water)
5. 3301170 A-56-A Diaphragm w/A-29—12 per pkg.
6. Brass Guide N/A; use 3301236. NOTE: 3301236 A-163-A Guide replaces all previous guides.
7. Inside parts N/A; see item no. 7 listed with new style valve. Repair Kit replaces all previous inside parts.
8. Inside Parts N/A; see item no. 8 listed with new style valve. Repair Kit replaces all previous inside parts.
9. 3301176 A-56-AA Washer set—6 per pkg.
10. B-32-A CP Handle Assem. N/A; use 3302279.
11. A-6 CP Handle Coupling N/A; use 0301082.
12. B-7 CP Socket N/A; use 0302019.
13. B-32 CP Grip N/A; use 3302274.
14. Handle Repair Kit N/A; use 3302305.
15. C-42-A 3'' CP Push-Button Assem. N/A; use 0303351.
16. 3303347 3'' CP Push-Button Replacement Kit.
17. 3'' Push-Button Repair Kit N/A; use 3303396.
18. Spud Coupling Assem. CP N/A; see item no. 19 listed with new style valve.
19. Outlet Coupling Assem. CP N/A; see item no. 20 listed with new style valve.
20. V-100-A & V-100-AA Vacuum Breaker N/A; consult local plumbing wholesaler for proper V-500-A or V-500-AA Vacuum Breaker replacement.
21. H-550 CP Stop Coupling N/A; use 0308676.
22. * 0308801 H-551-A CP Adj. Tail 2 $1/16$'' long
23. H-540-A Series Control Stops N/A; see Control Stop Repair Kits or item no. 24 listed with new style valve for complete replacement.
24. H-6 CP Stop Coupling N/A; use 0308063.
25. * 0308026 H-5 CP Ground Joint Tail 1$1/4$'' long
26. H-545-AG Series Control Stops N/A; see Control Stop Repair Kits or item no. 27 listed with new style valve for complete replacement.

Fig. 6-17 (*continued*)

Roughing and Repair Information 77

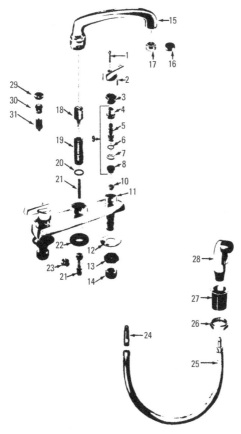

Fig. 6-18 Top-mount Aquaseal sink fitting with Hermitage trim (with parts list). *(Courtesy American Standard)* (*continued*)

1. Handle Screw
2. Handle
3. Locknut
4. Stem Nut
5. Stem w/Swivel
6. Friction Ring
7. Stop Ring
8. Aquaseal Diaphragm
9. Aquaseal Trim
10. Seat
11. Body
12. Friction Washer
13. Locknut
14. Coupling Nut
15. Spout
16. End Trim
17. Aerator
18. Diverter
19. Post
20. O-Ring
21. Hose Connection Tube
22. Gasket
23. Body Plug
24. Hose Connector
25. Hose S/A
26. Locknut
27. Spray Holder
28. Spray Head
29. Cap w/Washer
30. Auto Spray (Upper)
31. Auto Spray (Lower)

Note: If part #18 is not used, order parts 29, 30, & 31.

Fig. 6-18 (*continued*)

or groove is discovered, the seat should be replaced. This seat can be removed by using an Allen wrench in most cases.

If the seat is not too badly worn, and if the seat is the unremovable type, it can be refaced by using a *seat-dressing tool*. The cause of badly worn seats in most

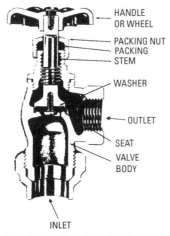

Fig. 6-19 Drawing of a globe valve.

cases is delay in replacing worn washers. The passing of water when the valve is shut between washer and seat causes a notch or groove to be worn.

7. When the stem is again inserted into the faucet body, remember to ensure that it is kept open slightly. This will prevent damage to the stem.

If water leaks out of the handle, it is caused by a worn O-ring, a thin rubber ring located on the stem. On some faucets, and on valves such as the globe valve illustrated in Fig. 6-19, the packing nut is the cause of leakage.

Open the valve (Fig. 6-15) a quarter-turn and tighten the nut snugly. If the valve continues to leak, a new packing washer must be installed. You may also wrap stranded graphite packing around the spindle and tighten snugly. If the spout (15) leaks where the body of the valve, this is also caused by a worn-out O-ring.

80 Chapter 6

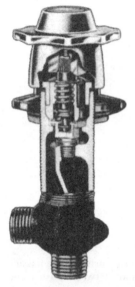

Fig. 6-20 No-drip Aquaseal valve.

The American Standard Aquaseal valve shown in Fig. 6-20 has a diaphragm (as illustrated in the Aquaseal kit, shown in Fig. 6-21) in place of a washer.

Faucets

The secret behind the no-drip feature of faucets lies in the washerless valve. All moving parts are outside the flow area, and lubrication on the stem threads is effective for the life of the fitting.

There is no seat washer wear, which is the cause of leaks and dripping in ordinary fittings.

Roughing and Repair Information 81

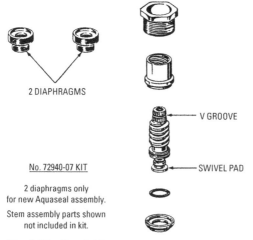

2 DIAPHRAGMS

No. 72940-07 KIT

2 diaphragms only for new Aquaseal assembly.

Stem assembly parts shown not included in kit.

V GROOVE

SWIVEL PAD

Fig. 6-21 Repair kit.

When it becomes necessary to replace the diaphragm in an Aquaseal valve, remove the handle and check for a V groove around the stem, located in the middle of the splines. If you *do not see* the V groove, replace the handle and contact your supplier for the appropriate stem assembly. If the V groove is visible, proceed to remove the valve unit.

After removing the old diaphragm, turn the stem so that one thread still protrudes from the top of the stem nut. Slip a new diaphragm over the swivel pad and insert the assembly into the fitting, exercising normal care to prevent damaging the diaphragm. Tighten the valve unit and replace the handle.

Fig. 6-22 shows an exploded view of the American Standard single-control Aquarian sink fitting.

82 Chapter 6

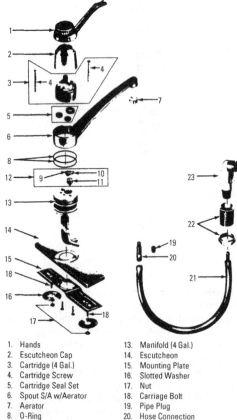

1. Hands
2. Escutcheon Cap
3. Cartridge (4 Gal.)
4. Cartridge Screw
5. Cartridge Seal Set
6. Spout S/A w/Aerator
7. Aerator
8. O-Ring
9. Diverter
10. O-Ring
11. Retainer S/A
12. Diverter & Retainer Set
13. Manifold (4 Gal.)
14. Escutcheon
15. Mounting Plate
16. Slotted Washer
17. Nut
18. Carriage Bolt
19. Pipe Plug
20. Hose Connection
21. Hose S/A
22. Locknut & Spray Holder
23. Spray Head S/A

Fig. 6-22 Single-control Aquarian sink fitting.

Roughing and Repair Information **83**

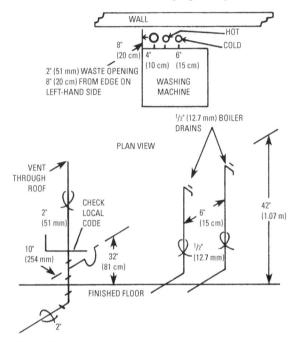

Fig. 6-23 Rough-in specifications for a washing machine.

Domestic Washing Machines

The drawings in Fig. 6-23 show rough-in specifications and a schematic diagram of water and waste lines running to and from a washer. Be sure, of course, to check local codes.

Two-Lever Handle Service Sink Faucet, Wall-Mounted with Pail Hook

The two-handle compression wall-mounted faucet shown in Fig. 6-24 has an electroplated rough chrome cast brass valve

Fig. 6-24 Two-handle service sink faucet.
(Courtesy Eljer Plumbingware Co.)

body. The faucet can be adjusted $7\frac{3}{4}$ inches to $8\frac{1}{4}$ inches. The spout swivels and has hose threads. The spout extends 4 inches from the wall. The wall-to-ball hook on the faucet is 9 inches. A chrome-plated vacuum breaker is part of the assembly (see Fig. 6-25). It has $\frac{1}{2}$-inch FIP (IPS) swivel unions with $\pm \frac{1}{4}$-inch offset. The faucet is designed to provide 5 or more gallons per minute at 80 psi. Note the faucet has a pail hook for holding buckets while they fill up.

Installing Kitchen Faucets

Kitchen faucets and those in bathrooms are often changed to fit the new homeowner's decorating taste. They are designed with the do-it-yourselfer in mind (see Fig. 6-26 for required

Roughing and Repair Information 85

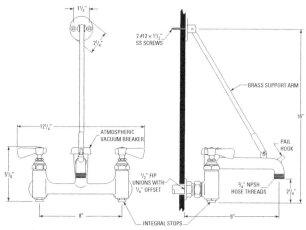

Fig. 6-25 Dimensions to aid in mounting the service sink faucet.
(Courtesy Eljer Plumbingware Co.)

tools). The steps in Fig. 6-27 show how easy it is to put in new faucets. However, after a few leaks and straining to reach things under the sink, the professional plumber is often called to finish the job.

Fig. 6-26 Tools needed to install a kitchen faucet.
(Courtesy Eljer Plumbingware Co.)

1. Shut off hot and cold water supply under sink. Plug sink drain with cloth to avoid losing small parts. Remove old faucet.

2. Place plastic plate on bottom of faucet, smooth side down. For models without a spray, proceed to step 5.

Refer to this if unit is a single-handle faucet

5. Set faucet in place on sink. From underneath sink, put washers and locknuts on finger-tight. Position faucet on sink properly and tighten nuts with basin wrench or adjustable wrench. Using a basin wrench makes this job easier.

Refer to this if unit is a two-handle faucet

3. Place spray hose guide in far right hole of sink and attach washer and locknut from underneath sink.

4. Insert threaded end of spray hose down through hose guide; attach to faucet. Pass hose through center hole in sealing gasket and attach to center faucet outlet, tightening securely. **Note:** Use pipe joint compound or plumbers' tape on this connection.

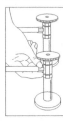

6. Note: If faucet supply lines need to be bent, grip tube at point of bend with right hand, and firmly and slowly pull tube into position using rounded bend with left hand. Extreme care must be exercised. Follow instructions exactly. Kinked tubes will void warranty.

7. Connect hot and cold water supplies. **Note:** If additional adapters or supply tubes are required, you can purchase them from the store where you bought the faucet. Use pipe compound or plumbers' tape when attaching faucet to water lines.

8. Turn on hot and cold water. Check for any leaks, especially at hookups to supply lines. Remove aerator from end of spout. Flush both water lines, allowing water to run for 1 minute. Replace aerator.

Fig. 6-27 Step-by-step instructions on installing the kitchen faucets. *(Courtesy Eljer Plumbingware Co.)*

Installing Single-Handle Tub and Shower Set

This system must be set by the installer to ensure safe, maximum temperature. Tools needed for this task are shown in Fig. 6-28. Any change in the setting may raise the discharge temperature above the limit considered safe and may lead to hot-water burns (see Fig. 6-29). It is the responsibility of the installer to properly install and adjust this valve.

This valve does not automatically adjust for inlet temperature changes. Therefore, someone must make the necessary temperature-limit stop adjustments at the time of installation, and further adjustments may be necessary because of a seasonal water temperature change. You must inform the owner or user of this requirement. According to industry standards, the maximum allowable water discharge temperature is 120°F. This temperature may vary in local areas. The typical temperature for a comfortable bath or shower is between 90° and 110°F.

Fig. 6-28 Tools needed to install single-handle tub and shower set. *(Courtesy Eljer Plumbingware Co.)*

Installing a Whirlpool Bathtub

The installation of a whirlpool bathtub is somewhat more complicated than a regular tub. The water and electrical aspects both must be considered. Suitable clearance and access are critical to proper installation. There are a number of important items to consider before getting started. Keep in mind

1. Shut off hot- and cold-water supply lines and remove old fittings, if any. Make opening in back wall large enough to work through.

2. Connect water supply lines—hot water to left and cold water to right. NOTE: Use Teflon tape or pipe compound on all threaded connections. When sweating a valve, all plastic and rubber components should be removed from the casting to protect them from the heat.

3. Install 1/2" shower supply line (not included) to center connection on valve assembly and to a 1/2" pipe elbow (not included) at desired shower height. For shower installation only, plug bottom outlet.

4. Assemble 1/2" tub spout supply line to 1/2" elbow (not included). Screw this assembly into bottom opening of valve assembly. Pipe elbow should reach wall opening just above tub. For tub installation only, plug top outlet.

5. Apply pipe compound or Teflon tape to tub spout nipple (not included). Attach tub spout to elbow. Tighten by hand.

6. Place flange over long end of shower arm and apply pipe compound or Teflon tape to threads on end. Screw into elbow at the shower supply line. The shower head can now be attached to shower arm. CAUTION: Use cloth over shower head nut to prevent scratching during installation.

7. Slip escutcheon over valve cover and screw into wall. Slip acrylic handle onto stem and screw securely into place using the screws provided. Snap on index button. Adjust the temperature limit stop; see CAUTION statement in step 6.

8. To flush water lines, let faucet run fully open through tub spout (unless shower only) for 1 minute each in the hot and cold positions.

Fig. 6-29 Step-by-step instructions on installing the single-handle tub and shower set.
(Courtesy Eljer Plumbingware Co.)

that the example used here is an Eljer tub. There may be differences in other manufacturers' products that require closer scrutiny.

Important
Read complete instructions before beginning installation. Fig. 6-30 shows the end view and side view along with the location of the drain/overflow. The specifications for the rectangular tub are shown in Table 6-1.

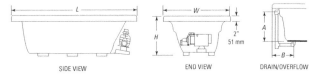

Fig. 6-30 Views of the rectangular whirlpool tub.
(Courtesy Eljer Plumbingware Co.)

Each whirlpool bath arrives ready for installation, completely equipped with motor/pump assembly, as well as plumbing and fittings necessary for whirlpool operation. An optional drain/overflow kit is available for installation on the bath.

Remove the bath from the carton. Retain the shipping carton until satisfactory inspection of the product has been made. *Do not lift the bath by the plumbing at any time; handle by the shell only.*

Immediately upon receipt, inspect the shell before installing. Should inspection reveal any damage or defect in the finish, do not install the bath. Damage or a defect to the finish claimed after the bath is installed is excluded from the warranty. Eljer's responsibility for shipping damage ceases upon delivery of the products in good order to the carrier. Refer any claims for damage to the carrier. For definitions

Table 6-1 Rectangular Tub Specifications

Model	Dimensions (Length × Width × Height)	Drain/Overflow Dimensions	Cutout	Total Weight/Floor Loading	Operating Gallonage	Product Weight	Skirt and Mounting
Cypress (5 feet by 32 inches)	60 inches (1524 mm) × 32 inches (813 mm) × 19¼ inches (489 mm)	15¾ inches (400 mm)/ 9¼ inches (235 mm)	58 inches × 30 inches	604 lbs (274 kg)/45 lbs per ft² (220 kg per ft²)	43 U.S. gallons (163 liters)	79 lbs (36 kg)	Optional
Savoy (5 feet by 42 inches)	60 inches (1524 mm) × 42 inches (1067 mm) × 18¼ inches (464 mm)	14⅜ inches (365 mm)/ 8¾ inches (222 mm)	58 inches × 40 inches	789 lbs (358 kg)/45 lbs per ft² (220 kg per ft²)	46 U.S. gallons (174 liters)	97 lbs (44 kg)	Optional
Patriot (6 feet by 36 inches)	72 inches (1829 mm) × 36 inches (914 mm) × 19¼ inches (489 mm)	15⅝ inches (400 mm)/ 9½ inches (241mm)	70 inches × 34 inches	745 lbs (338 kg)/ 42 lbs per ft² (205 kg per ft²)	52 U.S. gallons (197 liters)	97 lbs (43 kg)	Optional

Model	Dimensions		Weight	Capacity	Shipping Weight	Heater
Emblem (6 feet by 42 inches)	72 inches (1829 mm) × 42 inches (1067 mm) × 20½ inches (521 mm)	15¾ inches (400 mm)/8½ inches (216 mm)	70 inches × 34 inches	885 lbs (401 kg) / 42 lbs per ft² (205 kg per ft²)	63 U.S. gallons (239 liters)	99 lbs (45 kg)	Optional
Berkeley (5 feet by 36 inches)	60 inches (1524 mm) × 36 inches (914 mm) × 19¼ inches (489 mm)	15¾ inches (400 mm)/ 9¼ inches (235 mm)	58 inches × 34 inches	643 lbs (292 kg) / 43 lbs per ft² (210 kg per ft²)	49 U.S. gallons (186 liters)	85 lbs (39 kg)	Optional
Dakota (6 feet by 48 inches)	72 inches (1829 mm) × 48 inches (1219 mm) × 18¼ inches (464 mm)	14½ inches (368 mm)/ 11⅜ inches (238 mm)	70 inches × 46 inches	885 lbs (401 kg)/ 37 lbs per (181 kg ft²)	74 U.S. gallons (280 liters)	118 lbs (54 kg)	Optional

For all units:

Motor/pump: 115 VAC, 3450 rpm/7.8 amps, 60 Hz, single-phase.
Electrical requirements: 115 VAC, 15 amps, 60 Hz. Requires dedicated circuit. (Courtesy Eljer Plumbingware Co.)

92 Chapter 6

of warranty coverage and limitations, refer to the published warranty information packed with the product.

All bath units are factory-tested for proper operation and watertight connections prior to shipping.

Note
Prior to installation, the bath must be filled with water and operated to check for leaks that may have resulted from shipping damage or mishandling. The manufacturer is not responsible for any defect that could have been discovered, repaired, or avoided by following this inspection and testing procedure.

Fig. 6-31 shows rimless oval baths in the side view and end view and the location of the drain/overflow. Table 6-2 gives the specifications for the oval tubs.

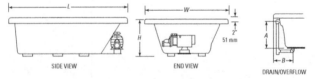

Fig. 6-31 Views of the rimless oval bathtub.

Corner baths are shown in Fig. 6-32 with the specifications of the corner tubs given in Table 6-3.

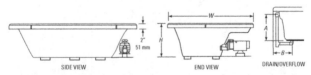

Fig. 6-32 Views of the corner bathtub.

Table 6-2 Oval Tub Specifications

Model	Dimensions (Length × Width × Height)	Drain/Overflow Dimensions	Cutout	Total Weight/Floor Loading	Operating Gallonage	Product Weight	Skirt and Mounting
Laguna 5 (5-foot oval)	62 inches (1575 mm) × 43 inches (476 mm) × 18¾ inches (476 mm)	15⅛ inches (384 mm)/ 9½ inches (241 mm)	Template provided	788 lbs (358 kg)/ 58 lbs per ft² (283 kg per ft²)	50 U.S. gal (189 liters)	96 lbs (44 kg)	Not available
Laguna 6 (6-foot oval)	72 inches (1829 mm) × 42 inches (1067 mm) × 20½ inches (521 mm)	16 inches (406 mm)/ 11⅜ inches (289 mm) per ft²	Template provided	880 lbs (400 kg)/ 52 lbs (254 kg per ft²)	68 U.S. gal (258 liters)	105 lbs (48 kg)	Not available

For all units:
Motor/pump: 115 VAC, 3450 rpm/7.8 amps, 60 Hz, single-phase.
Electrical requirements: 115 VAC, 15 amps, 60 Hz. Requires dedicated circuit.
(Courtesy Eljer Plumbingware Co.)

Table 6–3 Corner Tub Specifications

Model	Dimensions (Length × Width × Height)	Drain/Overflow Dimensions	Cutout	Total Weight/ Floor Loading	Operating Gallonage	Product Weight	Skirt and Mounting
Triangle (5-foot corner, 5-foot DE, Esquina)	60 inches (1524 mm) × 60 inches (1524 mm) × 19¾ inches (502 mm)	14⅞ inches (378 mm)/11 inches (279 mm)	Template provided	770 lbs (350 kg)/ 41 lbs per ft² (2.54 kg per ft²)	47 U.S. gal (178 liters)	119 lbs (54 kg)	Optional
Triangle II (60 inches × 60 inches)	60 inches (1524 mm) × 60 inches (1524 mm) × 22 inches (559 mm)	17¾ inches (451 mm)/ 10⅞ inches (276 mm)	Template provided	872 lbs (396 kg)/ 40 lbs per ft² (195 kg per ft²)	62 U.S. gal (235 liters)	122 lbs (55 kg)	Not available

For all units:
Motor/pump: 115 VAC, 3450 rpm/7.8 amps, 60 Hz, single-phase.
Electrical requirements: 115 VAC, 15 amps, 60 Hz. Requires dedicated circuit.
(Courtesy Eljer Plumbingware Co.)

Roughing and Repair Information   95

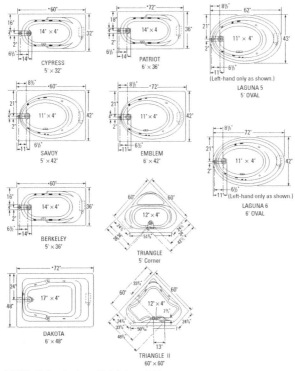

* Add 1/4" to this dimension when roughing-in for three-wall niche.
Note: 1. Measurements inside each unit represent cutout in floor to allow for drain/overflow.
      2. All measurements are in inches. To convert to millimeters, multiply inches by 25.4.

Fig. 6-33 Roughing-in references for whirlpool tubs.

Fig. 6-33 gives the roughing-in references for the various models of whirlpool tubs manufactured by Eljer. Note that unless otherwise specified, the units are produced with left- or right-hand version.

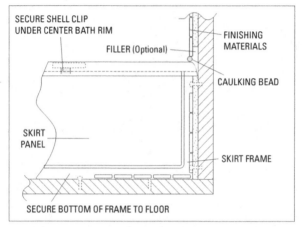

Fig. 6-34 U-frame skirt mounting detail.

A U-frame skirt mounting detail is shown in Fig. 6-34.

Service Access
For partially or fully sunken installations, allow for access to service connections. It is the installer's responsibility to provide sufficient service access. The recommended minimum dimensions allowable for service to the bath are shown in Fig. 6-35 and Fig. 6-36.

Provide adequate area around the unit for air circulation for cooling the motor and to supply sufficient air to the jets. Do not insulate this area or around motor.

Service Access with Skirt
An optional skirt fits along the side of the bath for above-floor installations and is also an access panel for servicing. Allow a space of at least 8 inches away from the bath for skirt removal.

Roughing and Repair Information **97**

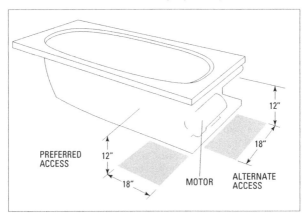

Fig. 6-35 Service access (without the skirt).

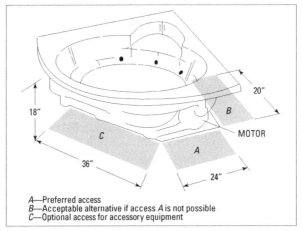

A—Preferred access
B—Acceptable alternative if access *A* is not possible
C—Optional access for accessory equipment

Fig. 6-36 Service access (triangle 5-foot corner).

98 Chapter 6

The skirt is designed to accommodate the added height of the tile, linoleum, or other floor coverings up to $1\frac{1}{4}$ inches above the floor and will be flush with the floor when installed.

More detailed instructions on skirt installation are provided with the optional skirt assembly.

Electrical Connections

A separate circuit, which must be protected by a ground fault circuit interrupter (GFCI), is required. Install a duplex outlet to the stud wall underneath the bathtub, at least 4 inches above the floor (see Fig. 6-37 and Fig. 6-38). The duplex outlet is not provided. Because these units are manufactured with a safe, convenient on-off switch on the bath itself, a remote switch or timer is not necessary. If an optional timer is desired, a switch for 120 volts AC operation can be used.

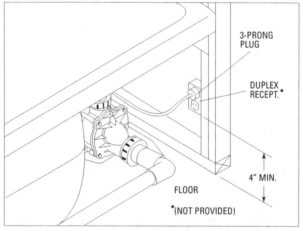

Fig. 6-37 Electrical connection for side/end drain baths.

Roughing and Repair Information 99

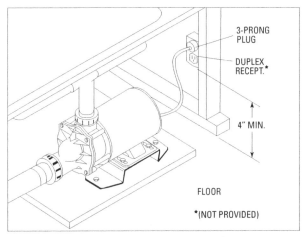

Fig. 6-38 Electrical connection for corner bath.

Danger

There is a risk of electrical shock. Connect only to a circuit protected by a GFCI.

Caution

By operating the motor/pump without enough water in the bath, you can cause leaking and permanent damage to the pump. Before power is applied to the installation, make sure the switch is in the "Off" position to avoid pump damage.

Drain Information

A drain/overflow assembly (sold separately) must be installed on the bath, water-tested, and connected to the sanitary system of the house. After opening the carton, inspect for damage and verify that the kit is of the proper finish. In the drain/

overflow kit, note that the waste flange, strainer, overflow cover, and cover screws are packaged in a separate package within the kit to protect the trim finish. Follow the installation instructions provided with the drain/overflow kit. After the drain is fully installed, test for proper drainage. If the unit does not drain correctly, rectify this condition before proceeding with the installation. The manufacturer is not responsible for removal and/or reinstallation costs.

Note
Watertight installation of the drain is the installer's responsibility. Drain leakage is excluded from the warranty of this product.

Plumbing
The pump, jets, and suction fittings for the whirlpool system are factory-plumbed in Schedule 40 PVC piping.

All Eljer products are factory-tested for proper operation and watertight connections prior to shipping. If leaks are detected, notify your dealer. Do not install the unit.

Water Supply
Consult local authorities for plumbing code requirements in your area.

Important
Proper installation of the fill spout plumbing and compliance with local codes are the responsibility of the installer. The manufacturer does not warrant connections of water supply fittings and piping, fill systems, or drain/overflow systems, nor is it responsible for damage to the bath that occurs during installation.

Caution
A nonflammable protective barrier must be placed between soldering work and bath unit to prevent damage to the bath.

Cleanup After Installation

To avoid dulling and scratching the surface of the bath, never use abrasive cleaners. A mild liquid detergent and warm water will clean soiled surfaces.

Remove spilled plaster with a wood or plastic edge. Metal tools will scratch the surface. Spots left by plaster or grout can be removed if lightly rubbed with detergent on a damp cloth or sponge.

Paint, tar, or other difficult stains can be removed with paint thinner, turpentine, or isopropyl alcohol (rubbing alcohol). Minor scratches that do not penetrate the color finish can be removed by lightly sanding with 600-grit wet/dry sandpaper. You can restore the glossy finish to the acrylic surface of the bath with a special compound, Meguiar's #10 Mirror Glaze. If that is not available, use automotive rubbing compound followed by an application of automotive paste wax.

Major scratches and gouges that penetrate the acrylic surface will require refinishing. Ask your local dealer for special instructions.

Operation

All baths manufactured by Eljer are designed for fill and drain, which means the bath should be drained after each use and filled with fresh water by the next bather. This is a health precaution because these baths are not designed to hold water continuously like pools or spas.

Once the bath is installed, remove any residue or foreign materials left over from construction. Use turpentine or paint thinner to remove stubborn stains, paint, or tar. Dirt can be cleaned off with a mild liquid detergent on a damp cloth. Scrape off plaster with a wooden or plastic edge. *Do not use metal scrapers, wire brushes, or other metal tools, because they will damage the bath's surface.*

102 Chapter 6

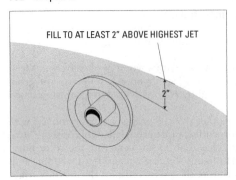

Fig. 6-39 Fill level for proper operation.

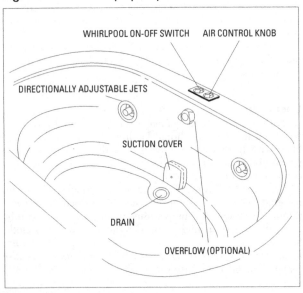

Fig. 6-40 Typical bath fittings.

Water Level

Close the drain and fill the bath until water is at least 2 inches above the highest jet (see Fig. 6-39). Do not turn on the whirlpool system at any time if the jets are not completely immersed in water. Running the whirlpool system when there is insufficient water in the bath could result in water spraying outside the bath area. Running the whirlpool system without water will damage the recirculating pump.

Whirlpool Switch

The whirlpool on-off switch, conveniently located on the bath, as shown in Fig. 6-40, allows the whirlpool system to be turned on and off while in the bath. Simply push down on the switch button to turn on the whirlpool system. To turn the system off, push down on the button again (see Fig. 6-41).

ON-OFF SWITCH

DEPRESS CENTER ON/OFF

Fig. 6-41 On-off switch mounted on rim of bathtub.

If your bath has an optional wall-mounted timer, set the time you wish the whirlpool to operate.

Note

When you desire less than 10 minutes of whirlpool action, it is necessary to turn the timer knob clockwise past the number 10 and then back to the desired amount. If the whirlpool action does not begin when the timer is correctly set, it is necessary to push the switch button.

Controlling Whirlpool Action

The whirlpool action in your bath is influenced by three factors—direction of flow, force of water, and force of air. All baths manufactured by Eljer are equipped with directionally

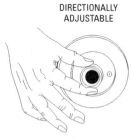

Fig. 6-42 Adjusting the direction of the jets.

adjustable jets that can be adjusted for direction and flow of air (see Fig. 6-42).

To change the direction of the water flow, swivel the jet nozzle to the desired angle. The jets can be directed individually toward any location on your body to provide a hydro-massage. The jets can also be adjusted so that they all point in the same direction (clockwise or counterclockwise) to circulate the water in a circular motion around the bath, causing a total whirlpool effect.

Fig. 6-43 Air induction controls. To increase airflow, turn counterclockwise. To reduce airflow, turn clockwise.

Two knobs located on the bath serve as controls for the air induction system. The intensity of the hydro-massage whirlpool action is determined by the amount of air inducted into the water. As the amount of air is increased, the hydro-massage action increases. For maximum air induction, rotate the control knobs fully counterclockwise to the largest circles. For fewer air bubbles, decrease the amount of air induction by rotating the control knob clockwise. When the knobs are turned to the smallest circles, only water is being circulated, as shown in Fig. 6-43.

7. OUTSIDE SEWAGE LIFT STATION

Pumping stations are built when sewage must be raised from a low point to a point of higher elevation, or where the topography prevents downhill gravity flow. Special nonclogging pumps are available to handle raw sewage. They are installed in structures called *lift stations*.

There are two basic types of lift station: wet well and dry well. A *wet-well installation* has only one chamber or tank to receive and hold the sewage until it is pumped out. Specially designed submersible pumps and motors can be located at the bottom of the chamber, completely below the water level. *Dry-well installations* have two separate chambers, one to receive the wastewater and one to enclose and protect the pumps and controls. The protective dry chamber allows easy access for inspection and maintenance. All sewage lift stations, whether of the wet-well or dry-well types, should include at least two pumps. One pump can operate while the other is removed for repair.

This chapter provides information intended to serve as a guide to familiarize the plumber or fitter with the most important features of a typical sewage lift station installation.

Outside Sewage Lift Station Piping

Fig. 7-1 shows a cutaway drawing of an outside sewage lift station. Fig. 7-2 shows the discharge piping extending out through the top of the ejector. Fig. 7-3 illustrates the piping arrangement when the discharge piping comes from the side of the basin. Fig. 7-4 shows a cross-sectional view of the pump with reference numbers for construction features. Fig. 7-5 is a cross-sectional reference drawing for guide bearing. Fig. 7-6 is a reference drawing for adjusting the impeller in the pump featured herein (should it ever become necessary). Fig. 7-7 shows the mercury float switch, which is used in conjunction with duplex pump applications.

106 Chapter 7

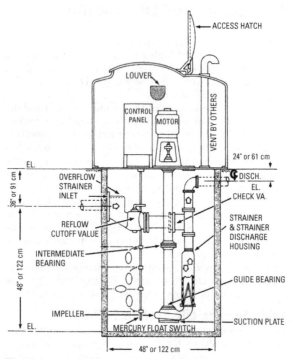

Fig. 7-1 Sewage lift station.

Fig. 7-8 shows the fiberglass outside pump housing. This is a lightweight fiberglass, recommended for outdoor installations, and is constructed of high-strength polyester resin reinforced with glass fibers. Durable weatherproof construction is ensured, and painting is not necessary. Because of the

Outside Sewage Lift Station **107**

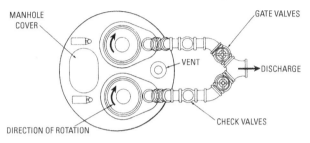

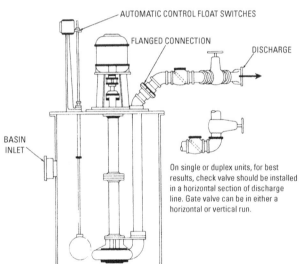

On single or duplex units, for best results, check valve should be installed in a horizontal section of discharge line. Gate valve can be in either a horizontal or vertical run.

Fig. 7-2 Discharge piping layout.

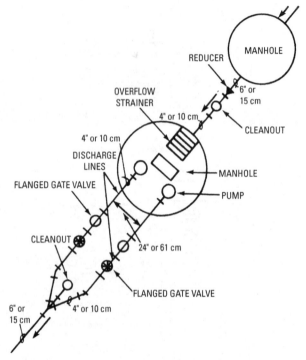

Fig. 7-3 Piping discharging from side of basin.

Outside Sewage Lift Station 109

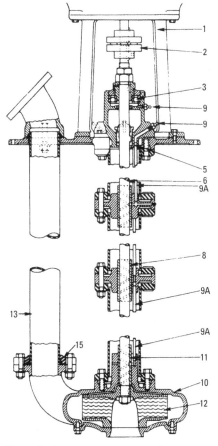

Fig. 7-4 Cross-sectional view of pump.

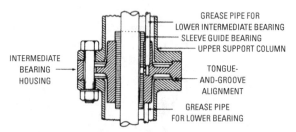

Fig. 7-5 Cross-sectional reference diagram.

housing's light weight, it is very easy to install or remove when it becomes necessary for a major pump overhaul.

General Instructions for Vertical Pumps

If installed, vertical centrifugal pumping units are very simple to maintain.

Before installing these pumps in a basin, make sure that the basin is fairly clean.

Any accumulation of sand, dirt, cinders, and so on, should be cleaned out, or unnecessary wear to the pump will occur. If solid matter is allowed to accumulate, it will gradually close off the suction of the pump. It is far more expensive to replace worn parts than to clean the basin at regular intervals.

As shown in Fig. 7-6, the impeller must be held in the center of the space provided for it in the pump casing and must not rub against the casing. Turn the pump shaft by hand. If the shaft does not turn freely, that is an indication that the impeller is rubbing. A micrometer adjustment is provided at the ball thrust bearing to raise or lower the shaft and impeller to the proper position. Do not change this adjustment unless it is necessary.

Outside Sewage Lift Station 111

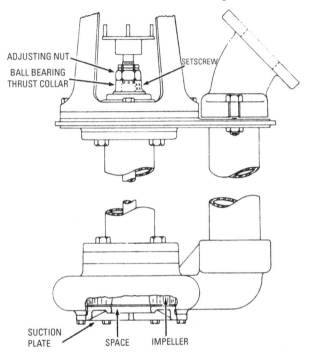

Fig. 7-6 Reference diagram for adjusting impeller.

To make an adjustment, loosen the setscrew in the ball bearing thrust collar and back off the adjusting nut slightly. This nut is a combination adjusting and locknut. It fits tightly around the shaft threads and will offer some resistance in turning. When backing off the nut, turn the shaft, pressing

112 Chapter 7

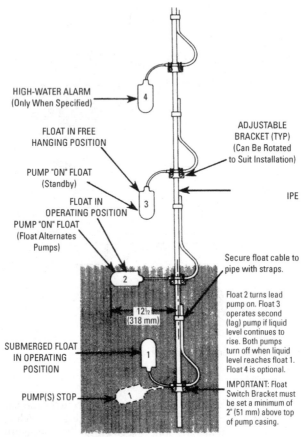

Fig. 7-7 Mercury float switch.

Outside Sewage Lift Station **113**

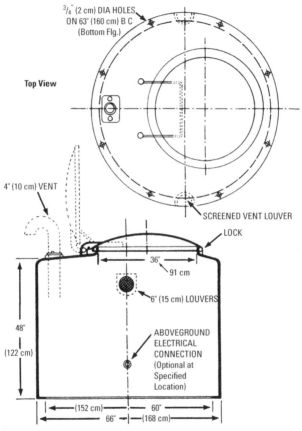

Fig. 7-8 Outside pump housing.

downward so that the impeller rubs on the suction plate. Take up the nut until the impeller just clears the suction plate and the shaft turns freely. Retighten the setscrew in the thrust collar.

Before turning on the current, be sure the shaft rotates freely. Check the direction of rotation (see the arrow on the pump floor plate).

For ball bearings and guide bearings, it is a good idea to add 2–3 ounces (57–85 grams) of grease at a time, at regular intervals, until it can be determined how often more grease will be needed. The grease tubes for lubricating the lower guide bearings are enclosed in the hanger pipes.

Pump Construction Features

All joints in the pump suspension system are tongue-and-groove type. The intermediate bearings and bearing housings are self-contained assemblies that ensure ease of maintenance and proper fit and alignment after dismantling. Following are some important pump features:

- *Motor support*—Top end machined to match the NEMA C motor end flange. No shims are necessary for perfect alignment.
- *Flexible coupling*—Properly sized and designed for the pump load and the motor speed.
- *Ball thrust bearing*—Located in a sealed housing and protected from dirt. It is of ample capacity to carry the weight of the pump shaft and impeller.
- *Stuffing box*—Contains packing.
- *Pump shaft*—This must be of ample diameter in all pump sizes to prevent any whipping action. It provides a large factor of safety for handling maximum loads and shocks in pumping unscreened sewage.

- *Guide bearing*—This provides stability to the extended shaft to keep it aligned and rotating easily.
- *Pressure grease fittings*—For all bearings.
- *Lube pipe*—Located in the shaft support column, except where a large number of intermediate bearings in a pump more than 12 feet (3.66 m) long makes it necessary to locate it outside.
- *Pump casing*—This is cast iron joined to the steel suspension pipe with a tongue-and-groove flange for rigid, permanent alignment. Note the wide, smooth passages permitting free flow at all points.
- *Casing bearing*—This is located close to the impeller hub to reduce overhang. It is designed for maximum wearing surface and equipped with a forced-feed grease line for oil lubrication as specified. Forced grease is recommended.
- *Impeller*—This is the most important single part of the sewage pump. Here is where experience counts, and it is important to obtain pumps with the best-quality impellers. The impeller is tight fitting, held in rotation with a stainless steel shaft key, and dynamically balanced.
- *Discharge pipe*—This leads away from the pumping area and conducts the effluent to the sewer.
- *Expansion fitting*—Prevents distortion and strain on the shaft and bearings.

Wastewater Collection Systems

A wide variation in sewage flow rates occurs over the course of a day. A sewer system must accommodate this variation. In most cities, domestic sewage flow rates are highest in the morning and evening hours. They are lowest during the middle of the night. Flow quantities depend on the population

116 Chapter 7

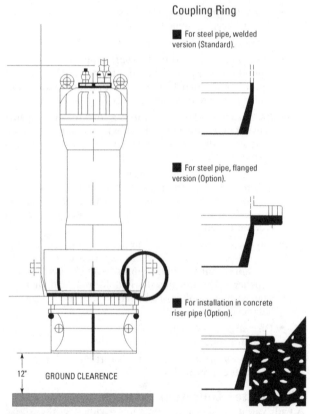

Fig. 7-9 Submersible pump for sewage and wastewater applications.

Outside Sewage Lift Station **117**

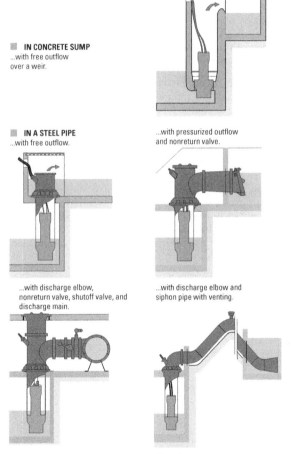

■ IN CONCRETE SUMP
...with free outflow
over a weir.

■ IN A STEEL PIPE
...with free outflow.

...with pressurized outflow
and nonreturn valve.

...with discharge elbow,
nonreturn valve, shutoff valve, and
discharge main.

...with discharge elbow and
siphon pipe with venting.

Fig. 7-10 Various installations for the submersible pump.

density, water consumption, and the extent of commercial or industrial activity in the community. The average sewage flow rate is usually about the same as the average water use in the community. In a lateral sewer, short-term peak flow rates can be roughly four times the average flow rate. In a trunk sewer, peak flow rates may be two-and-a-half times the average.

Submersible Axial Flow Pumps

Submersible axial flow pumps (see Fig. 7-9) are for direct installation in the discharge pipes. These save space and installation costs. They are intended for large flows and moderate heads for storm water, land drainage, and flood protection. The axial flow pump can handle sewage from commercial, municipal, and industrial sources. Its semi-open impellers are designed for trouble-free pumping of liquids containing solids and fibrous material.

The compact units are lowered into standard steel tubes and need no anchoring (see Fig. 7-10). Their own weight is sufficient to hold the units securely in position. The robust design ensures high operational reliability and hydraulic efficiency up to 88 percent.

8. PIPES AND PIPELINES

This chapter provides an overview of the installation of pipes and pipelines, particularly pipelines hung from a ceiling, connections between a heater and storage tank, and shock absorbers for hot- and cold-water supply lines.

Pipelines Hung from a Ceiling

Instead of running pipe in a ceiling, where possible, sleeve locations should be transferred from the floor above to the floor below. Make use of columns and walls to square off your work. Once points to be reached are established, and you know in which direction you are heading and where you are coming from, it will become clear what fittings will work. Cuts can then be determined, and even the location of hangers can also be determined, if inserts were not provided in the building construction.

The next step is transferring these hanger locations to the ceiling by way of your plumb bob.

When running hot- and cold-water headers to a number of fixtures in the same area (especially a battery of lavatories), the headers should be run as shown in Figs. 8-1, 8-2, and 8-3. Water supply pipes extending vertically one or more stories are called *risers*. Soil and vent pipes extending vertically are called *stacks*. A 2 percent grade is slightly less than $1/4$ inch per foot.

For example, consider a sewer 220 feet (37 meters) long:

$$\begin{array}{r} 220 \\ \times .02 \\ \hline 4.40' \end{array}$$ or approximately $52^{3/4}''$—total fall

120 Chapter 8

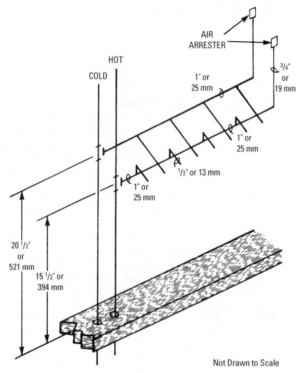

Fig. 8-1 Hot- and cold-water headers for lavatories.

In metric—a sewer 37 meters long:

$$\begin{array}{r} 37 \\ \underline{.02} \\ .74 \end{array}$$

.74 meters, or 740 mm—total fall in 37 meters on 2% grade

Pipes and Pipelines 121

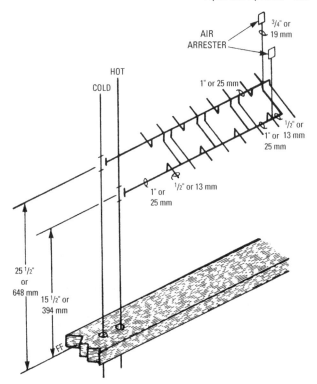

Fig. 8-2 Hot- and cold-water headers for back-to-back lavatories.

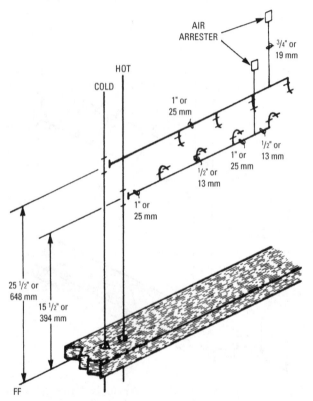

Fig. 8-3 Schematic of hot- and cold-water headers.

When a line is neither horizontal nor vertical, it is said to be slanting or diagonal.

A 1-inch (25-mm) water pipe is equal to four $1/_2$-inch (13-mm) pipes.

Eighteen inches (457.2 mm) of 4-inch (101.6-mm) pipe holds 1 gallon (3.7854 liters) of water.

Pipe more than 12 inches (305 mm) in diameter is generally classified by its outside diameter (OD). Thus, 14-inch (355.6-mm) pipe has an OD of 14 inches.

A *cross-connection* is any physical connection or arrangement of piping that provides a connection between a safe water supply system and a separate system (or a source that is unsafe or of questionable safety) and (under certain conditions) permits a flow of water between the safe and unsafe systems or sources.

To be effective, *air chambers* should be located as close as possible to the points at which *water hammer* will occur. Water hammer becomes much greater at 100 psi (689.5 kPa) than at 50 psi (345 kPa). Two types of manufactured devices used to reduce or eliminate water hammer are *shock absorbers* or *water hammer arrestors* to cushion the shock and *pressure-reducing valves* to lower the operating pressure (see Fig. 8-4).

Condensation is formed on cold-water pipelines when warm, humid air comes into contact with the cold surfaces, causing these pipelines to give up some of their moisture.

A *corporation stop* is located at the tap in the city water main. The connection between the city water main and the building is called the *service pipe*.

The *invert* of a sewer line or any pipeline is the inside-wall flow line at the bottom of the pipe.

Brass pipe expands about $1^1/_4$ inches (32 mm) per 100 feet (30.5 meters) for a 100°F (38°C) rise in temperature.

Steel pipe expands about $3/_4$ inch (19 mm) per 100 feet (30.5 meters) for a 100°F (38°C) rise in temperature.

124 Chapter 8

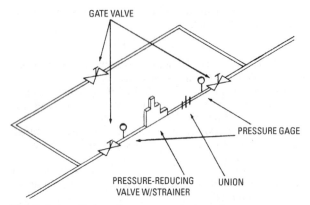

Fig. 8-4 Installation of pressure-reducing valve.

When pressure exceeds 80 psi (552 kPa), a pressure regulator should be installed, especially where such pressure leads to a hot-water heater. Check the local code.

Connections Between Heater and Storage Tank with Bypass

Fig. 8-5 shows connections between a heater and storage tank with a bypass.

Hot- and Cold-Water Supply Line Shock Absorbers

Most plumbing codes dictate that all water supply and distribution systems shall be provided with air chambers or other approved mechanical devices to suppress water hammer line noises and to prevent pressure hazards to the piping system. Mechanical water suppressors shall be accessible for inspection and repair. Fig. 8-6 illustrates typical shock absorber

Pipes and Pipelines 125

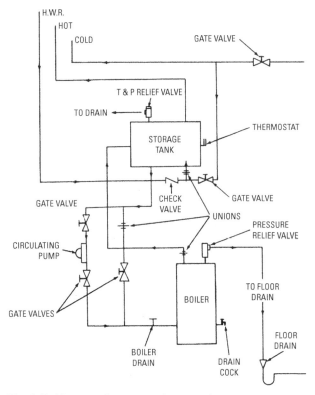

Fig. 8-5 Heater and storage tank connections.

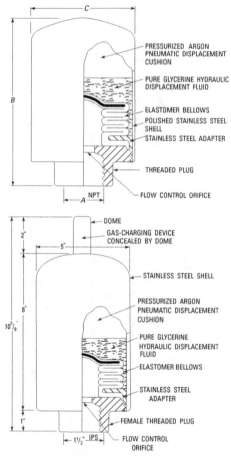

Fig. 8-6 Hot- and cold-water supply line shock absorbers.

Pipes and Pipelines 127

Table 8-1 Table of Dimensions for Hot- and Cold-Water Supply Line Shock Absorbers

Fixture Units	Dimensions			(Weight) (Approximate)
	A	B	C	
1–11	$^3/_4$ inch	$4^1/_2$ inches	$3^1/_4$ inches	2 lbs 12 oz
12–32	1 inch	$5^1/_4$ inches	$3^1/_4$ inches	3 lbs 4 oz
33–60	1 inch	6 inches	$3^1/_4$ inches	3 lbs 6 oz
61–113	1 inch	$6^3/_4$ inches	$3^1/_4$ inches	3 lbs 8 oz
114–154	1 inch	$6^3/_4$ inches	5 inches	7 lbs 5 oz
155–330	1 inch	$7^3/_4$ inches	5 inches	8 lbs 2 oz

devices Table 8-1 shows the dimensions for these units. Fig. 8-7 illustrates a typical location for these units in hot- and cold-water supply branches.

It is recommended that valve controls within a building be provided for each hot-water tank, water closet, urinal, and lawn sprinkler. These valves are not always required in single-family dwellings. However, in public buildings or multiple-dwelling units, a valve is recommended for control at the base of each riser and for each dwelling unit or public toilet

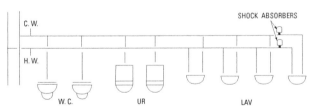

Fig. 8-7 Typical location of shock absorbers in a water line.

room, unless served by an independent riser. Control valves are also recommended for installation of fixtures isolated from a group. A main shutoff valve shall be installed near the street curb and gutter line on the exterior of the building.

Many residences have a short pipe soldered or chemically welded to the supply line of a sink, water closet, clothes washer, or other device. The short (usually 12 inches long) piece of pipe will have a cap soldered to the end, trapping air inside, which is then compressed whenever the water pressure starts to fluctuate.

9. VENTS, DRAIN LINES, AND SEPTIC SYSTEMS

This chapter provides background information on vents, discusses the functionality of venting in a drainage system, and examines some examples of common venting. The discussion concludes with a look at septic systems.

Definitions

A *continuous* vent (also called a *back vent*) is a vertical vent that is a continuation of the drain to which it connects. The *main* vent (also called a *vent stack*) is the principal artery of the venting system to which vent branches may be connected. A *branch* vent is a vent pipe connecting one or more individual vents with a vent stack or stack vent. A *wet* vent is a waste pipe that also serves as an air-circulating pipe or vent.

A *circuit* vent is a branch vent that serves two or more traps and extends from in front of the last fixture connection of a horizontal branch to the vent stack. An *individual* vent is a vent pipe installed to vent a fixture trap. It may connect with another vent pipe 42 inches (1.07 m) or higher above the fixture served, or it may terminate through the roof individually. A *dual* vent (also called *common* or *unit* vent) is a vent connecting at the junction of two fixture drains and acting as a vent for both fixtures. A *relief* or *yoke* vent is a vent where the main function is to provide circulation of air between a drainage and vent system.

A *local* vent is a ventilating pipe on the fixture inlet side of the trap. This vent permits vapor or foul air to be removed from a fixture or room. This removal of foul air or offensive odors from toilet rooms is accomplished now by bathroom ventilation fans and ducts. A *dry* vent conducts air and vapor only to the open air. A *loop vent* is the same as a circuit vent, except that it loops back and connects with a stack vent instead of a vent stack.

Purpose of Venting in a Drainage System

The purpose of venting is to provide equal pressure in a plumbing system. Venting prevents pressures from building up and retarding flow. It protects trap seals and carries off foul air, gases, and vapors that would form corrosive acids harmful to piping.

A plumbing system is designed for not more than 1 inch (2.5 cm) of pressure at the fixture trap. Greater pressures may disturb the trap seals. One inch (2.5 cm) of pressure is equivalent to that of a 1-inch column of water.

A *waste stack* terminates at the highest connection from a fixture. From this point to its terminal above the roof, it is known as a stack vent. In colder climates, the closer the vent terminal is to the roof, the less chance of frost closure.

Examples of Venting

Figs. 9-1 through 9-11 illustrate a number of venting situations. The system in Fig. 9-1 delivers fresh air for an ice machine, steam table, vegetable sink, and powder rooms in a public building. Fig. 9-2 shows a vent and drainage system in a residence. Figs. 9-3, 9-4, 9-5, 9-6, and 9-7 show vent systems for water closets, tubs, and lavatories. Fig. 9-8 shows combination and vent stacks, Fig. 9-9 shows water closets in a looped vent system, and Fig. 9-10 illustrates a looped vent with a bleeder. In all cases, walls and partitions have been eliminated for simplicity. Fig. 9-11 shows a multibath venting system.

Where sewage-ejector and fresh-air vents are used, both must extend through the roof independently. A garbage disposal must empty directly into the sanitary system, not the fresh-air system. The horizontal branch waste line should be 3 inches (76.2 mm) and continue up to the sink opening.

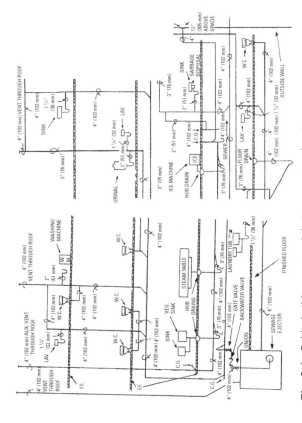

Fig. 9-1 Public building vent and drainage system drawing.

132 Chapter 9

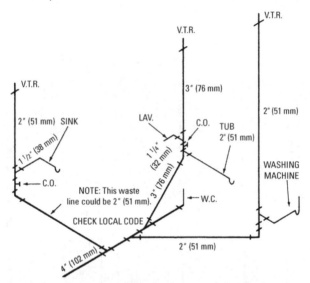

Fig. 9-2 Residential vent and drainage system drawing.

Check local code in all cases. The following common abbreviations are used in the drawings: C.O. (cleanout), W.C. (water closet;), and V.T.R. (vent through roof).

If a garbage disposal is installed on a sink, the horizontal branch waste line should be 3 inches (76.2 mm) and continue up to the sink opening. Check local code.

Septic Tanks

A septic tank will work properly for many years if installed correctly. A septic tank should be made watertight and airtight so that the bacterial action that disintegrates the solid matter can take place (see Fig. 9-12).

Vents, Drain Lines, and Septic Systems 133

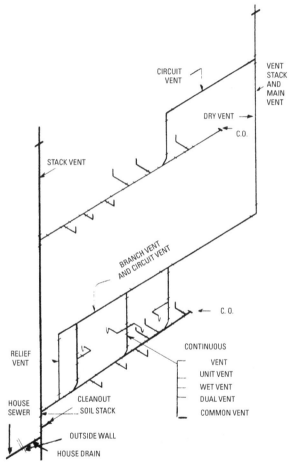

Fig. 9-3 Drawing of water closet and lavatory vents.

134 Chapter 9

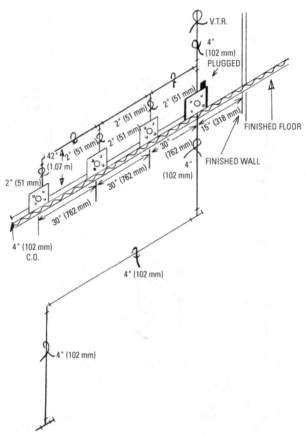

Fig. 9-4 Venting system for wall-hung water closets.

Vents, Drain Lines, and Septic Systems **135**

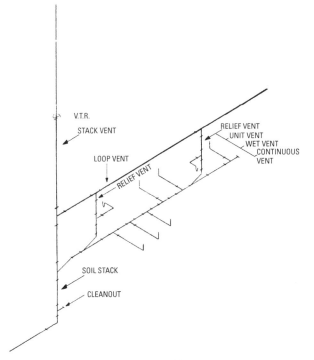

Fig. 9-5 Another water closet and lavatory venting system.

Septic tanks are designed to operate full. While the bacterial action works on the solids, the liquid overflow drains off into the drain field.

The tank itself does not need a cleanout, since the precast slabs can easily be removed if it ever becomes necessary to

136 Chapter 9

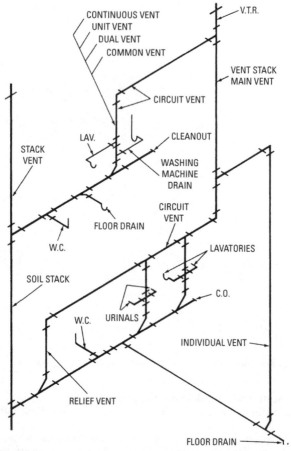

Fig. 9-6 Venting system for wall-hung urinals, floor drains, and water closets.

Vents, Drain Lines, and Septic Systems **137**

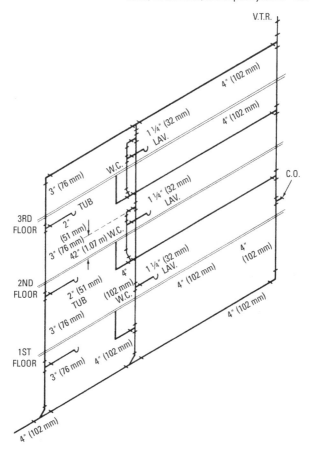

Fig. 9-7 Tubs, water closets, and lavatories are vented in this system.

138 Chapter 9

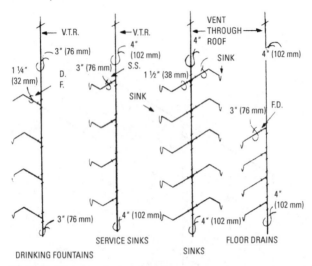

Fig. 9-8 System using combination waste and vent stacks.

clean the tank. Once a tank is cleaned, and precast slabs are set back into place, they can then be resealed with cement or mortar mix minus sand or gravel.

Drain fields can and will become spent in time, and new drain lines must be laid. These new lines must begin at the distribution box and extend out in new directions.

Information may vary according to location and type of soil. Check local code.

The following data is for septic systems in hard compact soil:

- Two-bedroom house—750-gallon (2839-liter) capacity with 200 feet (61 m) of drain field.

Vents, Drain Lines, and Septic Systems **139**

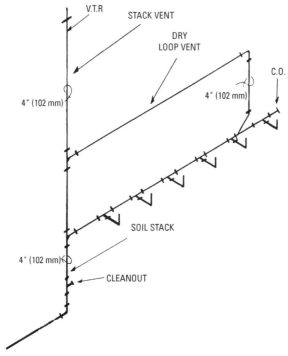

Fig. 9-9 Water closets in a looped vent system.

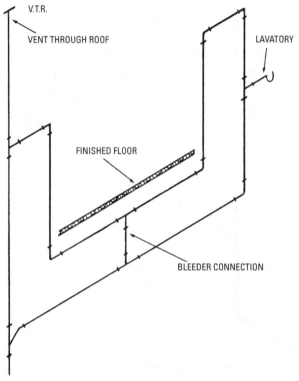

Fig. 9-10 Looped vent using a bleeder.

Vents, Drain Lines, and Septic Systems 141

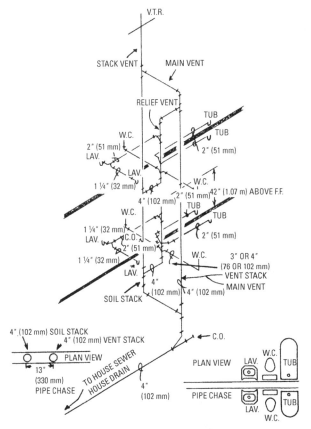

Fig. 9-11 Multibath venting system.

142 Chapter 9

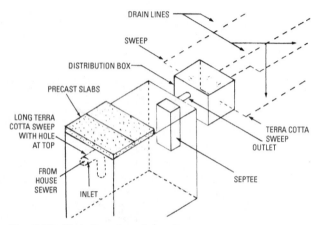

Fig. 9-12 Typical septic tank installation.

- Three-bedroom house—1000-gallon (3785-liter) capacity with 300 feet (91.5 m) of drain field.
- Four-bedroom house—1000-gallon (3785-liter) capacity with 400 feet (122 m) of drain field.

In sandy soil, smaller tanks and less drain field footage are generally the rule.

The top of the tank is generally 6 inches (152.4 mm) to 10 inches (254 mm) below ground level. The distance from the inlet invert (or bottom part of the inlet) to the septic tank from the top of the tank is generally 12 inches (305 mm). The outlet invert is generally 2 inches (51 mm) lower.

If drain lines must run parallel to each other, these lines should be at least 10 feet (3.05 m) apart. Terra cotta or cement drain tile is used, measuring 4 inches (101.6 mm) inside and 12 inches (304.8 mm) long.

Beginning Septic Tank Drain Lines

After the tank is set, you begin the drain field by placing one drain tile between the outlet opening and the distribution box.

At each opening leading to a drain line, a ditch about 24 inches (609.6 mm) wide is dug 6 inches (152.4 mm) below the outlet opening of the distribution box. Wooden pegs are then driven in the ground, beginning at each outlet of the box and spaced every 12 inches to 18 inches (304.8 mm to 457.2 mm) apart. The top of the peg should be level with the bottom of the outlet openings in the distribution box. Pegs should be laid level, or pitched approximately 1 inch (25.4 mm) every 100 feet (30.48 m).

Next, the crushed rock or gravel is installed to the level of the pegs or 6 inches (152.4 mm).

The drain tile is laid next, with care taken to space each tile approximately $^3/_8$ inch (10 mm) apart. These cracks or spaces are to be covered with tar paper.

After the tile is set, each drain line will then receive more crushed rock until rock reaches 1 inch (25.4 mm) above the drain tile.

The last step is to cover the entire drain line with tar paper and then cover with earth.

Note

A 1000-gallon (3785-liter) septic tank contains approximately 133.75 cubic feet (3.787 cubic meters) of space. This size tank could measure 42 inches (1066.8 mm) wide × 84 inches (2133.6 mm) long × 66 inches (1676.4 mm) deep.

A 750-gallon (2839-liter) septic system contains approximately 100.25 cubic feet (2.838 cubic meters) of space. This size tank could measure 35 inches (889 mm) wide × 72 inches (1828.8 mm) long × 68.75 inches (1746.25 mm) deep.

Measurements are inside dimensions.

10. LEAD WORK

This chapter provides general tips and recommended procedures for plumbers working with lead piping.

Working with Wiping Solder
A plumber should buy only good wiping solder that has the manufacturer's name and brand indicating composition cast on the bar. Refusal to accept anything less will eliminate many solder troubles that can and do occur. Good wiping solder usually contains between 37 percent and 40 percent pure tin, with between 63 percent and 60 percent pure lead.

Complete solidification occurs around 360°F (182°C); complete liquefaction occurs around 460°F (238°C). In general, the working range will be about 100°F (38°C).

Many plumbers twist a piece of newspaper and dip it into molten solder to check for correct wiping heat. When the paper scorches but does not ignite, the solder is hot enough.

Preparing Horizontal Round Joints
First, the ends of the pipe to be joined should be squared off with a coarse file or rasp and the pipe should be drifted so that it is uniformly round and free from dents.

Grease and dirt should be cleaned off the surface of the pipe for about 4 inches (102 mm) from the ends. With a knife, ream out burrs in the ends of the pipe to be joined.

Then, flare one end (using a turnpin and mallet) until the inside diameter at the end equals the original outside diameter. The end thus flared is the one in which the water will flow. The shoulder or outside edge of the flare should then be rasped off (see Fig. 10-1) approximately parallel with the outside wall of the pipe. An adequate flare for lead pipe is $1/4$ inch (6.4 mm) to $3/8$ inch (9.5 mm).

The inside of the flared end should be soiled for about 1 inch (25 mm). Next, the end of the other section of pipe

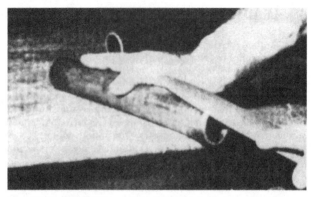

Fig. 10-1 Beveling the male end of lead pipe with a rasp preparatory to fitting it into the flared end for wiping. A close fit is highly important to successful joint wiping.

(which will be the one from which the water will flow) should be beveled with a rasp until it fits snugly inside the flared end. In this way, the joint is made in the direction of the flow, reducing resistance and chance of clogging. With dividers, mark a line around the flared end of the pipe, at a distance from the extreme end, equal to half the length of the finished joint (half being generally 1¼ inches, or 32 mm).

Mark a similar line on the beveled end at a distance from the extreme end equal to half the length of the joint, plus the length of the beveling. This will make the center of the finished joint at the intersection of the two outside surfaces of pipe. Next, mark an additional 3 inches (76 mm); in this 3-inch (76-mm) portion (which will be at the extreme ends of both pieces of pipe to be joined), clean lightly with wire brush, dust off, and apply plumber's soil.

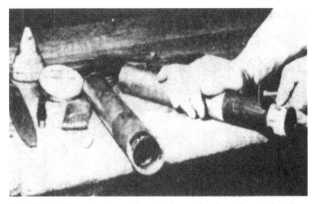

Fig. 10-2 Scraping the male end clean after beveling and soiling. Female end, at left, flared, soiled, and cleaned, is ready to have the male end inserted into it.

Next, the $1^{1}/_{4}$-inch (32-mm) portions on each pipe (totaling $2^{1}/_{2}$ inches, or 64 mm), the flare, and the bevel should be lightly scraped clean (see Fig. 10-2) with a shave hook and immediately covered with a thin coating of tallow to prevent oxidation.

Now, the ends of the pipe should be fitted snugly together and braced in position (see Fig. 10-3) so that they will be absolutely stationary during and after wiping until the solder has cooled.

The bottom of the pipe should be about 6 inches (152 mm) above the bench or working place. When wiping joints are in place, if the space is more than 6 inches (152 mm), a box or some other flat object should be placed so that there will be a surface about 6 inches (152 mm) under the joint to prevent splashing of solder.

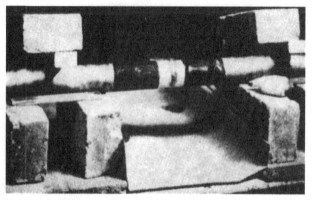

Fig. 10-3 Pipes prepared and fitted together ready for wiping. Note how they are held securely by boards and bricks with small boards under one side to prevent rolling. The paper under the joint is to catch excess solder.

To aid in getting heat up on a joint quickly, one or both of the outer or extreme ends of the pipe should be plugged (usually with newspaper). Proper heat is 600°F (316°C).

After the wiping solder is heated, carefully stirred and skimmed to remove dross, and tested for proper heat (as described previously), wiping may begin. Wiping cloths are usually made of 10-oz (284-gram) herringbone material. The wiping cloth sizes generally used are 3-inch (76-mm) cloth for a 2½-inch (64-mm) joint and 3¼-inch (83-mm) cloth for a 2¾-inch (70-mm) joint, measured by length of joints. Figs. 10-4 to 10-7 show solder being applied to a joint.

Procedure for Cleaning Wiping Solder

The procedure for cleaning wiping solder varies. One method is to heat solder to a dull red, about 790°F (421°C) (melting

Lead Work **149**

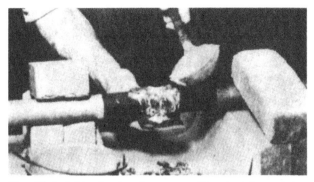

Fig. 10-4 Solder has just been poured on the joint and that caught in the cloth is being pressed against the bottom to get up heat.

Fig. 10-5 Excess solder has been removed from the soiling, and shaping has been started across the bottom and up the side next to the wiper.

150 Chapter 10

Fig. 10-6 The hand has been reversed and the stroke continues across the top and down the side away from the wiper.

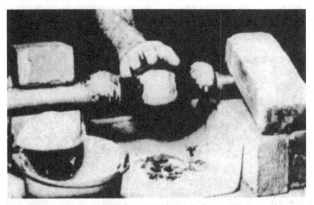

Fig. 10-7 A handy method of retaining under adverse conditions by building up the solder a little way from each side of the joint to be lifted off after completion.

point of zinc), and then add about 1 tablespoon of sulfur and stir. Then, permit the pot to cool slowly and skim off the top dross, which contains the impurities consisting of compounds of lead, tin, and zinc. Stir and skim until the top is clean. Next, add a small amount of powdered or lump rosin, stir, and skim again. Allow the pot to cool until the solder reaches wiping temperature. Then, add sufficient tin to reestablish proper workability.

Wiping Head Stub on Job Brass Ferrule

Prepare by using a fine file, apply Nokorode, and proceed to tin the ferrule.

For the lead stub, use a fine file to remove rough edges and a knife to ream. With a flat dresser, proceed to work the lead until it fits snugly into the brass ferrule (assume this type of ferrule is used). Approximately $1^{1}/_{8}$ inches (29 mm) to $1^{1}/_{4}$ inches (32 mm) should be inserted into the ferrule. Next, using the shave hook, shave a portion equal to 1 inch (25 mm) plus the insertion, say, $2^{1}/_{4}$ inches (57 mm), and apply mutton tallow. Then, insert into the ferrule and proceed to secure. Apply a ring of plumber's soil inside the ferrule brass side where the lead ends inside.

Next, use dividers and mark a 1-inch (25-mm) line around the lead from the face of the ferrule. Next, use a wire brush to clean an additional 3 inches (76 mm) of lead. Apply soil on this portion. Try not to apply to the 1-inch (25-mm) portion being readied to receive solder. Next, use a shave hook to clean a 1-inch (25-mm) portion. Then, reapply mutton tallow. Last, apply gummed paper to the ferrule. After wiping, prepare the lead cap, readying the stub for testing. Fig. 10-8 shows a completed installation.

Lead-Joining Work

In the regular run of lead-joining work, the plumber usually needs a pair of 2-pound (907-gram) soldering irons. For

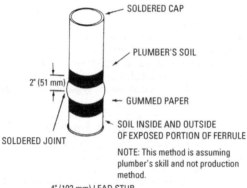

Fig. 10-8 Lead stub wiped on a job brass ferrule.

ordinary work, the point of the soldering iron should be tinned on all four sides for a distance of at least ³/₄ inch (19 mm) from the tip. If, however, soldering is to be done from underneath the joint, the iron should be tinned on one surface only (the side to be used next to the joint being soldered). This permits control of the solder and prevents it from running away from the joint.

Lap Joints

A lap of about ³/₈ inch (10 mm) is advisable for the weights of lead ordinarily employed by plumbers. On the top surface of the bottom sheet, a line should be marked ¹/₂ inch (12.7 mm) back from the edge to be joined. This portion should be shaved lightly with the shave hook, using strokes parallel to the edge, until the surface is clean back to the line.

The same procedure should be followed on the underside of the top sheet. The edge of the top sheet should also be cleaned and the top sheet placed in position, lapping over the other $3/8$ inch (10 mm). This leaves $1/8$ inch (3.2 mm) of the cleaned portion of the lower piece of lead exposed. With the flat dresser, the top sheet should be dressed down to the level of the bottom sheet, except where it actually laps and is held up by the lead underneath. The lap should be dressed to fit snugly. With the shave hook, the upper surface of the top layer should then be cleaned for a distance of $3/4$ inch (10 mm) from the edge. Tallow should be applied immediately to all cleaned areas. As in making butt joints, the sheets should be tacked in similar manner.

Butt Joints

To make a butt joint, the edges to be joined should be beveled with a shave hook so that they make an angle of 45° or more with the vertical. This is accomplished by wrapping a piece of cloth around the index finger of the hand holding the shave hook and using this finger, pressed against the edge of the lead, as a guide when drawing the shave hook along the edge.

Immediately after shaving, tallow candle or refined mutton tallow (free from salt) should be rubbed over all shaved parts in a very thin coat to prevent oxidation.

Edges to be joined should then be placed firmly together and powdered resin sprinkled along the joint. With a clean, well-tinned soldering iron and 50-50 solder, the edges are next tacked together at intervals of 4 inches (102 mm) to 6 inches (152 mm), using a drop of solder at each point. An iron at proper heat should then be placed against the lead at the end of the seam in the groove formed by the abutting beveled edges.

Solder should be fed in slowly, allowing it to be melted by the iron and fill the groove. The iron should be drawn slowly along the joint at a speed that permits the solder to melt and fill the groove continuously, building up to a slightly rounded surface when finished.

154 Chapter 10

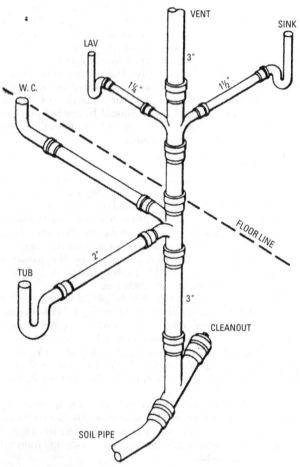

Fig. 10-9 A riser diagram showing stack venting for one bath group.

Riser Diagrams

Riser diagrams are elevations or perspective views showing stacks and risers, as displayed in Fig. 10-9. Generally, riser diagrams are shown only for complicated jobs. Plumbing drawing installation details are not generally placed on the plumbing drawings. Specifications indicate type and quality of fixtures, piping, and fittings. Trades workers will install materials in accordance with standard trade practices, drawings and specifications, and local code instructional requirements.

Soil Pipe

Soil pipe is commonly listed by standard length requirements (that is, by size, material, and strength or weight). Standard shipment is in 5-foot lengths. Fittings are classified by size, material, classification, weight or strength, and quantity.

In soil pipe, a 90° elbow is referred to as a quarter-bend, and a 45° elbow as an eighth-bend. Typical material is listed as follows:

- 4 inches extra-heavy cast-iron soil pipe (5-foot lengths) —10 lengths
- 4 inches eighth-bend extra-heavy cast iron—2 each
- 4 inches quarter-bend extra-heavy cast iron—4 each

Code Requirements

All plumbing codes require that joints and connections in the plumbing system be watertight and gastight. There are exceptions to watertight and gastight joints and connections. These exceptions include those sections of the systems that are open-joint or perforated piping and that have been installed for the purpose of collection and disposal of ground or seepage water to a below-surface storm drainage system.

11. LEAD AND OAKUM JOINTS

This chapter discusses important facts and techniques for plumbers to consider when working with lead and oakum joints.

Preparation

To prepare lead and oakum joints in pipe up to 6 inches (152 mm), the following tools are needed:

- 12-oz (0.340-kg) ballpeen hammer
- Caulking iron(s)
- Packing iron
- Yarning iron
- Joint runner (for pouring lead in horizontal joints)
- Wood chisel (steel handle) for cutting lead gate created by use of joint runner
- Chain snap cutters or ratchet cutters (for cutting pipe)

A cold chisel and a 16-oz (0.453-kg) ballpeen hammer can be used to cut pipe. Place a 2-inch × 4-inch (5-cm × 10-cm) piece of wood directly under the cut mark. Allow one end of the pipe to touch the ground, and with one foot placed near the desired cut, hold the pipe solidly against the wood.

Begin marking the pipe with light hammer blows. When the entire circumference of pipe is marked with cold-chisel indentation marks, begin using heavier blows.

Since pipe length is 5 feet, a 6-foot rule is usually called for in measuring. After the proper amount of oakum is placed in the joint, leaving 1 inch (25 mm) for lead (joints up to 6 inches or 152 mm), lead is then poured. The joint is now ready for caulking to make it watertight and airtight (see Fig. 11-1).

158 Chapter 11

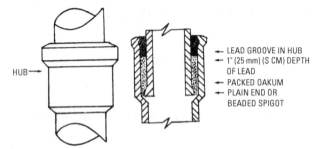

Fig. 11-1 Typical lead and oakum joint.

Note
Before dipping the ladle into the molten lead, be sure it is dry and free from moisture. Warm it over the lead pot while the lead is being heated. Moisture on a ladle will form steam when dipped into molten lead, causing an explosion.

Table 11-1 shows the relationship between pipe sizes, lead rings, oakum, and lead. Table 11-2 shows the information in the metric scale.

Caulk horizontal joints inside first. Outside first is preferred on vertical joints. When caulking, use moderate hammer blows. Each position of the caulking iron should slightly overlap the previous position. The lead and oakum joint provides a waterproof joint—strong, flexible, root-proof, watertight, and airtight.

Note
Approximately 8 lbs (3.62 kg) of brown oakum is used per 100 lbs (45.3 kg) of lead.

Six pounds (2.7 kg) of white oakum is used per 100 lbs (45.3 kg) of lead.

Caulking lead in cast-iron bell-and-spigot water mains should be 2 inches (51 mm) deep.

Table 11-1 Lead and Oakum Information

Size of Pipe and Fitting (Inch Joint)	Lead Ring Depth (inches)	White Oakum (ounces)	Brown Oakum (ounces)	Lead SV-Xh (lbs)
2	1	$1\frac{1}{2}$	$1\frac{3}{4}$	$1\frac{1}{4}$
3	1	$1\frac{3}{4}$	$2\frac{1}{2}$	$1\frac{3}{4}$
4	1	$2\frac{1}{4}$	3	$2\frac{1}{4}$
5	1	$2\frac{1}{2}$	$3\frac{3}{4}$	$2\frac{3}{4}$
6	1	$2\frac{3}{4}$	$3\frac{1}{2}$	3
8	$1\frac{1}{4}$	$5\frac{1}{4}$	7	6
10	$1\frac{1}{4}$	$6\frac{1}{2}$	$8\frac{1}{2}$	8
12	$1\frac{1}{4}$	$7\frac{1}{2}$	$9\frac{3}{4}$	$10\frac{1}{4}$
15	$1\frac{1}{2}$	$11\frac{1}{2}$	15	$17\frac{1}{4}$

Table 11-2 Lead and Oakum Information (Metric)

Size of Pipe and Fitting (mm)	Lead Ring Depth (mm)	White Oakum (grams)	Brown Oakum (grams)	Lead SV-Xh (grams)
51	25	43	50	567
76	25	50	71	794
102	25	64	85	1020
127	25	71	92	1247
152	25	78	99	1360
203	32	149	198	2721
254	32	184	241	3629
305	32	213	276	4876
381	38	326	425	8051

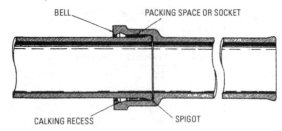

Fig. 11-2 A bell-and-spigot joint used for connecting lengths of cast-iron pipe.

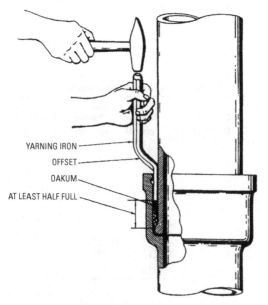

Fig. 11-3 Packing oakum into the bell-and-spigot of a sewer pipe.

Lead and Oakum Joints 161

Joints and Connections

All caulked joints of cast-iron bell-and-spigot soil pipe should be firmly packed with oakum or hemp. These should be secured only with pure lead to a minimum depth of 1 inch and should be run in one pouring and caulked tight (see Fig. 11-2). The bell-and-spigot joint shown in Fig. 11-2 is used for connecting lengths of cast-iron soil pipe. Fig. 11-3 illustrates a method used for packing oakum into the bell-and-spigot connection of a sewer pipe.

The use of jute, hemp, or oakum is not permitted for caulking for joining water distribution piping connections.

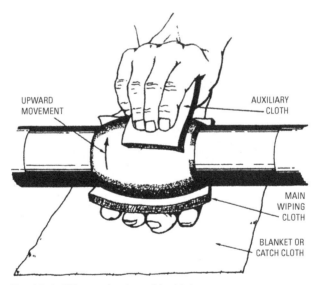

Fig. 11-4 Wiping a horizontal lead joint.

162 Chapter 11

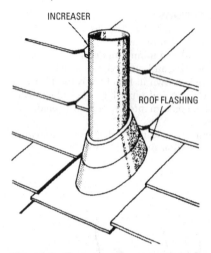

Fig. 11-5 Typical lead vent roof flashing.

Joints in lead pipe (or between lead pipe or fittings and brass or copper pipe, ferrules, soldering nipples, bushings, or traps) should be (in all instances) full-wiped joints. Exposed surfaces of each side of the wiped joints should be not less than ³/₄ inch and at least as thick as the material being joined. Fig. 11-4 shows a typical process of wiping a lead joint.

Fig. 11-5 shows the typical lead vent roof flashing.

12. SILVER BRAZING AND SOFT SOLDERING

This chapter provides an overview of silver brazing and soft soldering, which are used to ensure proper connections. Important safety tips for performing oxyacetylene welding and cutting appear at the end of the chapter.

Piping

Pipe is supplied in straight lengths in sections from 12 feet to 20 feet long. Standard piping of wrought iron or steel up to 12 inches in diameter is classified by its nominal inside diameter. The actual inside diameter may vary for a given nominal size (such as stand pipe, heavy pipe, or extra-heavy pipe), depending on pipe weight factors. The external diameter is normally the same for all three weights. Pipe larger than 12 inches is classified by its actual diameter.

Brass and copper piping are classified by the same nominal sizes as iron pipe with two weights for each size—extra strong and regular.

Fittings

Various lengths of pipe are connected to form a desired length or form by fittings that also provide continuity and direction changes. Pipe continuity is provided by the use of couplings, nipples, and reducers. Tees, crosses, and elbows are used to change direction of a run of piping. A *cap* is used to close an open pipe. A *plug* is used to close an open fitting. A *bushing* is used to reduce the size of an opening.

Unions provide convenient connections that can be easily unmade. Screwed unions are of three-piece design: two pieces are screwed to the ends of the pipes being connected, and the third draws them together by screwing onto the first piece and bearing against the shoulder of the second. Flow in a

Table 12-1 Brazing Information

Size of Copper Tube		Oxygen Pressure		Acetylene Pressure	
Inches	Metric	psi	kPa	psi	kPa
½ inch, ¾ inch	13 mm, 19 mm	5	34.5	5	34.5
1 inch, 1¼ inches	25 mm, 32 mm	6	41.4	6	41.4
1½ inches, 2 inches, 2½ inches	38 mm, 51 mm, 64 mm	7	48.26	7	48.26
3 inches, 3½ inches	76 mm, 89 mm	7½	51.7	7½	51.7
4 inches, 5 inches, 6 inches	102 mm, 127 mm, 152 mm	9	62	9	62

piping system is regulated by valves that are specified by type, material, size, and working pressures.

To ensure that the unions, tees, couplings, and valves are properly connected to copper pipe and some brass pipe and fixtures, it is necessary to either silver solder (silver brazing) or soft solder them. Keep in mind that acid core solder is used in wire form in many cases and solid sticks of solder with liquid flux in others. Of course, the plastic piping is welded together chemically and requires no heat.

Table 12-1 shows relevant brazing specifications for copper tubing.

Applying Heat and Brazing Alloy

The preferred method for brazing or silver soldering is by the oxyacetylene flame because a very high temperature is needed. Propane torches with portable tanks and other gases are usually used on smaller tube sizes. Silver solder needs a higher temperature (around 1725°F) to melt the silver solder. This is a very expensive way of soldering because of the required materials and is reserved for special soldering applications.

A slightly reduced flame should be used, with a slight feather on the inner blue cone. The outer portion of the flame is pale green. Heat the tube first, beginning at about 1 inch from the edge of the fitting. Sweep the flame around the tube in short strokes up and down at right angles to the run of the tube. It is very important that the flame be in continuous motion (and not be allowed to remain on any one point) to avoid burning through the tube.

Generally, the flux may be used as a guide to determine how long to heat the tube. Continue heating after the flux starts to bubble or work, and until the flux becomes quiet and transparent (like clear water). Flux has a tendency to clean the copper or brass and, at the same time, keep it from oxidizing at the hot spot. It allows the solder to flow easily (even upward).

Flux passes through four stages:

1. At 212°F (100°C) the water boils off.
2. At 600°F (316°C) the flux becomes white and slightly puffy and starts to work.
3. At 800°F (427°C) it lies against the surface and has a milky appearance.
4. At 1100°F (593°C) it is completely clear and active and has the appearance of water.

Avoid applying excess flux, and avoid getting flux on areas not cleaned. Particularly avoid getting flux into the inside of the tube itself (see Fig. 12-1 and Fig. 12-2). The purpose of flux is to dissolve residual traces of oxides, to prevent oxides from forming during heating, and to float out oxides ahead of the solder.

Now, switch the flame to the fitting at the base of the cup. Heat uniformly, sweeping the flame from fitting to tube until the flux on the fitting becomes quiet. Particularly avoid excessive heating of cast fittings.

166 Chapter 12

Fig. 12-1 Fluxing. *(Courtesy Copper Development Association)*

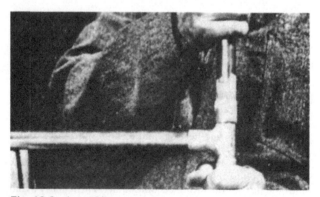

Fig. 12-2 Assembling. *(Courtesy Copper Development Association)*

Silver Brazing and Soft Soldering **167**

When the flux becomes liquid and transparent on both the tube and the fitting, start sweeping the flame back and forth along the axis of the joint to maintain heat on the parts to be joined, especially toward the base of the cup of the fitting. The flame must be kept moving to avoid burning the tube or fitting (see Figs. 12-3 through 12-6).

When the joint has reached proper temperature, apply brazing wire or rod where the pipe enters the fitting. Keep the flame away from the rod or wire as it is fed into joint. Keep both the fitting and the tube heated by moving the flame back and forth from one to the other as the alloy is drawn into the joint (see Fig. 12-7).

When the joint is filled, a continuous fillet of brazing alloy will be visible completely around the joint. Stop feeding as soon as the joint is filled.

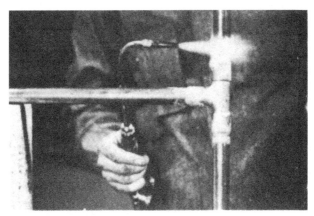

Fig. 12-3 Heating tube. *(Courtesy Copper Development Association)*

Fig. 12-4 Heating large tube. *(Courtesy Copper Development Association)*

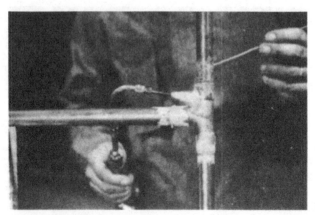

Fig. 12-5 Heating fitting. *(Courtesy Copper Development Association)*

Silver Brazing and Soft Soldering **169**

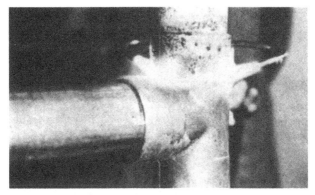

Fig. 12-6 Heating large fitting. *(Courtesy Copper Development Association)*

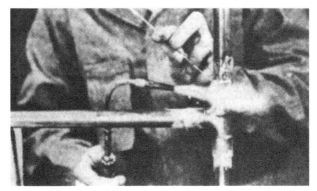

Fig. 12-7 Feeding brazing alloy. *(Courtesy Copper Development Association)*

Note
> For larger-size tube, 1 inch (25 mm) and above, it is difficult to bring the whole joint up to heat at one time.

If difficulty is encountered in getting the entire joint up to the desired temperature, a portion of the joint can be heated and brazed at a time. At the proper brazing temperature, the alloy is fed into the joint. The torch is then moved to an adjacent area. The operation is carried on progressively all around the joint, with care taken to overlap each operation.

Horizontal and Vertical Joints

When making horizontal joints, it is preferable to start applying the brazing alloy at the top, then the two sides, and finally the bottom, making sure that the operations overlap.

On vertical joints, it doesn't matter where the start is made. If the opening of the socket is pointed down, care should be taken to avoid overheating the tube. Overheating may cause the alloy to run down the tube. If that happens, take the heat away and allow the alloy to set. Then, reheat the back of the fitting to draw up the alloy (see Fig. 12-8). After the brazing alloy has set, clean off the remaining flux with a wet brush or swab (see Fig. 12-9).

Wrought fittings may be chilled quickly. However, it is advisable to allow cast fittings to cool naturally to some extent before applying a swab.

If the brazing alloy refuses to enter the joint and tends to flow over the outside of either member of the joint, it indicates this member is overheated, or the other is underheated, or both. If the alloy fails to flow or has a tendency to ball up, it indicates oxidation on the metal surfaces or insufficient heat on the parts to be joined.

Silver Brazing and Soft Soldering **171**

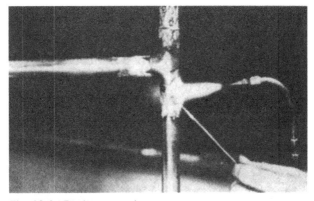

Fig. 12-8 Feeding upward. *(Courtesy Copper Development Association)*

Fig. 12-9 Swabbing. *(Courtesy Copper Development Association)*

Making Up a Joint

The preliminary steps of tube measuring, cutting, burr removing, and cleaning (tube ends and sockets must be thoroughly cleaned before beginning the brazing operation) are identical to the same steps in the soft-soldering process. Figs. 12-10 through 12-18 show this process.

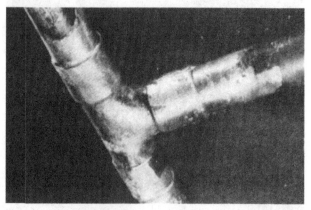

Fig. 12-10 Completed joint. *(Courtesy Copper Development Association)*

A flux can be made that will be suitable for making silver-solder joints on copper tubing by mixing powdered borax and alcohol or water to a thin, milky solution.

Soft Soldering

Keep the following in mind when performing soft soldering:

- Avoid pointing the flame into the socket opening of the fitting.
- Never apply the flame directly to solder. The tube being soldered should be hot enough to melt the solder.

Silver Brazing and Soft Soldering 173

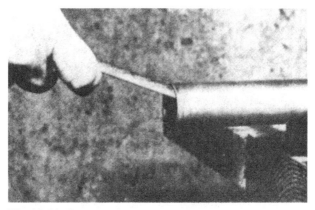

Fig. 12-11 Removing burrs. *(Courtesy Copper Development Association)*

Fig. 12-12 Cleaning tube end. *(Courtesy Copper Development Association)*

174 Chapter 12

Fig. 12-13 Cleaning fitting socket.
(Courtesy Copper Development Association)

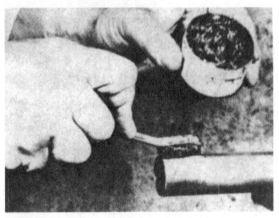

Fig. 12-14 Fluxing tube end. *(Courtesy Copper Development Association)*

Silver Brazing and Soft Soldering **175**

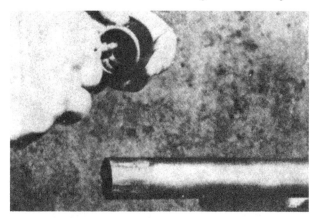

Fig. 12-15 Fluxing fitting socket.
(Courtesy Copper Development Association)

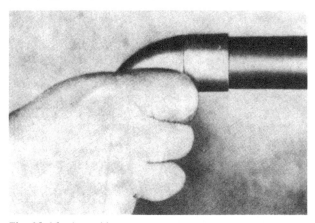

Fig. 12-16 Assembling. *(Courtesy Copper Development Association)*

Fig. 12-17 Removing surplus flux.
(Courtesy Copper Development Association)

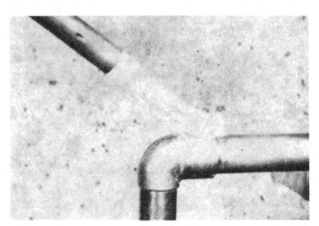

Fig. 12-18 Heating. *(Courtesy Copper Development Association)*

- On tubing 1 inch (2.5 mm) and above, a mild heating of the tube before playing the flame on the fitting is recommended. This results in a better-made joint, ensuring that solder is drawn into the joint by the natural force of capillary attraction.
- Flame should be played at the base of the fitting, with the flame pointing in the direction of the socket opening. This ensures that any impurities, including excess flux, will be flushed out ahead of the solder as the joint is filled.

  If the flame is pointed toward the base of the fitting, there is a chance of these impurities or flux being trapped inside the joint, creating a flux pocket. A *flux pocket* prevents solder from completely occupying the inside of the socket.
- When the material is hot enough, the flame should be moved away and the solder applied.
- On larger-size tubing, it is best to hold the flame long enough at the base of the fitting and then move it around the circumference to ensure evenly distributed heat and solder.
- As the joint cools, continue to apply solder around the entire face of the fitting. This will create a fillet that ensures a full joint. Heat rises, and sometimes the top part of a horizontal joint is too hot to retain solder, allowing it to run out (or in a tight, inaccessible place, a portion may not have been heated). This can be detected by running solder around the entire face. If it runs smoothly around, creating a filler, you have a better guarantee that a good joint has been made.
- On horizontal joints, it is recommended that the flame be played at the bottom of the fitting and solder applied

178 Chapter 12

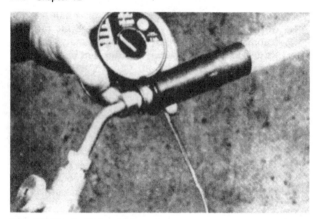

Fig. 12-19 Applying solder. *(Courtesy Copper Development Association)*

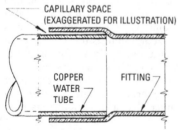

Fig. 12-20 A cross-sectional view of a joint positioned for soldering.

at the top; however, this is merely preferred by the majority of plumbers (see Fig. 12-19 and Fig. 12-20).

Basic Oxyacetylene Safety Measures

Safe practices to follow when oxyacetylene welding or cutting include the following:

- Always blow out cylinder valves before attaching the regulators. Dust can cause combustion, resulting in an explosion.

- Stand to the side of the regulator when opening the cylinder valve. The weakest point of every regulator is either front or back. The regulator could blow out, and an explosion could occur.

- Always release the adjusting screw on the regulator before opening the oxygen cylinder valve. When the adjusting screw is released, the seat of the regulator is in contact with the nozzle with sufficient pressure to hold the 2200 psi (15,168 kPa), so the oxygen released travels only a short distance.

    If the regulator were open when high pressure is released through the seat nozzle, there would be expansion going into the regulator and then restriction into the nozzle, thus generating considerable heat that could set off dust or oil.

- Always open the cylinder valve slowly. By opening the valve slowly, the heat generated from the travel is very small. The main reason is to reduce shock. If the valve is opened fast, the pressure exerted from the shock hitting the seat surface exceeds that of the pressure contained in the cylinder.

- A good practice is to light fuel gas before opening the oxygen valve on the torch. Light the large tip, show soot, then no soot, then flame leaving the tip. The burning rate should be set with the acetylene valve only. If you do open the oxygen valve first, you will pop the large tip.

- Never use oil on regulators, torches, and so on. Oxygen and oil create an explosion. In oxygen cylinders, there is as much as 2200 psi (15,168 kPa) of pressure. When the pressure is released from the cylinder through the regulator, the speed at which the oxygen travels exceeds the speed of sound, and this generates heat and friction. The smallest amount of oil (even just the oil from your skin) will ignite and blow up the regulator.
- Do not store cylinders near flammable material (especially oil, grease, or any other readily combustible substance).
- Acetylene cylinders should be stored in a dry, well-ventilated location.
- Acetylene cylinders should not be stored in close proximity to oxygen cylinders.
- Never tamper with safety devices in valves or cylinders. Keep sparks and flames away from acetylene cylinders, and under no circumstances allow a torch flame to come in contact with safety devices. Should the valve outlet of an acetylene cylinder become clogged by ice, thaw with warm (not boiling) water.
- Acetylene should never be used at a pressure exceeding 15 psi (103 kPa) gage.
- The wrench used for opening the cylinder valve should always be kept on the valve spindle when the cylinder is in use.
- Finally, points of suspected leakage should be tested by covering them with soapy water. *Never* test for leaks with an open flame.

Part II
Plumbing Systems

13. PLASTIC PIPE AND FITTINGS

Plastic drain, waste, and vent (DWV) piping has been approved by local and state codes including the Building Officials Conference of America, Southern Building Code Congress, International Association of Plumbing and Mechanical Officials, and Federal Housing Administration (FHA). Following are some types of plastic pipe:

- *Polyvinyl chloride (PVC)*—Type-1 polyvinyl chloride is strong, rigid, and economical. It resists a wide range of acids and bases but may be damaged by some solvents and chlorinated hydrocarbons. The maximum service temperature is 140°F (60°C). PVC is better suited to pressure piping.

- *Acrylonitrile-butadiene-styrene (ABS)*—Usage of ABS has almost doubled compared with PVC in DWV piping systems. However, it is limited to 160°F (71.1°C) water temperatures at lower pressures, which is considered adequate for DWV use.

- *Chlorinated polyvinyl chloride (CPVC)*—This meets national standards for piping 180°F (82.2°C) water at pressures of up to 100 psi (689 kPa). It can withstand 200°F (93.3°C) water temperature for limited periods. CPVC is similar to PVC in strength and overall chemical resistance.

- *Polyethylene (PE)*—This is a flexible pipe for pressure systems. Like PVC, it cannot be used for hot-water systems.

- *Polybutylene (PB)*—This is flexible and can be used for either hot-water or cold-water pressure systems. Since no method has been found to chemically bond PB, solvent-weld joints *cannot* be used. Compression-type joints are used instead.

- *Polypropylene*—This is a very lightweight material suitable for lower-pressure applications up to 180°F (82.2°C). It is used widely for industrial and laboratory drainage acids, bases, and many solvents.
- *Kem-Temp polyvinylidene fluoride (PVDF)*—This is a strong, tough, and abrasive-resistant fluorocarbon material. It has excellent chemical resistance to most acids, bases, and organic solvents and is ideally suited for handling wet or dry chloride, bromine, and other halogens. It can be used in temperatures of up to 280°F (138°C).
- *Fiberglass-reinforced plastic (FRP) epoxy*—This is a fiberglass-reinforced thermoset plastic with high strength and good chemical resistance up to 220°F (104.4°C).

Expansion in Plastic Piping

Following are some expansion characteristics of PVC pipe:

- *PVC Type 1*—100 feet (30.5 meters) operating at 140°F (60°C) will expand approximately 2 inches (50.8 mm).
- *CPVC*—Polypropylene and PVDF at the same temperature will expand approximately $3\frac{1}{4}$ inches (82.55 mm).

Applications

Plastic pressure piping for hot- and cold-water supply is now permitted in FHA-financed rehabilitation projects. Plastic pipe enjoys markets in natural gas distribution, rural potable water systems, crop irrigation, and chemical processing. Almost 100 percent of all mobile homes and travel trailers have plastic pipe.

Two types of plastic pipe and fittings are commonly used for drainage systems: PVC and ABS.

Plastic Pipe and Fittings 185

Joints with Plastic Tubing

Follow these steps when working with joints in plastic tubing:

1. Cut the tubing.
2. Test-fit the joint.
3. Apply primer (for PVC and CPVC only).
4. Apply cement.
5. Assemble the joint.
6. Allow to set.

Cutting the Tubing

It is important to use the right primer and/or solvent. Priming is essential with PVC and CPVC. No priming is needed with ABS. The *recommended practice* for making solvent-cemented joints with PVC and ABS pipe and fittings follows. Pipe should be cut square, using a fine-tooth handsaw and a miter box, or a fine-tooth power saw with a suitable guide (see Fig. 13-1). Regular pipe cutters may also be used (a special

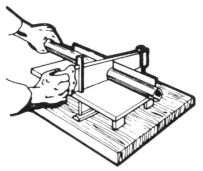

Fig. 13-1 Cut the pipe squarely. *(Courtesy NIBCO, Inc.)*

cutting wheel is available to fit standard cutters). Hand-held cutters resembling a pair of pliers or a hedge trimmer are now available for the smaller-diameter PVC tubing. All hardware stores carry them.

Great care should be taken to remove all burrs and ridges raised at the pipe end (see Fig. 13-2). If the ridge is not removed, cement in the fitting socket will be scraped from the surface on insertion, producing a dry joint and causing probable joint failure. All burrs should be removed with a knife, file, or abrasive paper.

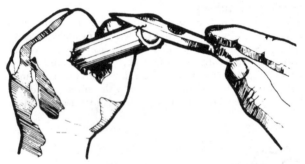

Fig. 13-2 Smooth the end of the pipe. *(Courtesy NIBCO, Inc.)*

Test-Fit the Joint

Wipe both the outside of the pipe and the socket of the fitting with a clean, dry cloth to remove foreign matter. Mate the two parts without forcing. Measure and mark the socket depth of the fitting on the outside of the pipe. Do not scratch or damage pipe surface to indicate where the pipe end will be bottomed.

The pipe should enter the fitting at least one-third (but not more than halfway) into the socket or fitting depth (see

Plastic Pipe and Fittings 187

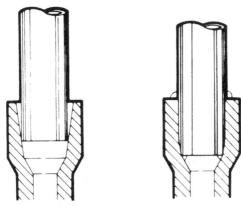

Fig. 13-3 Cross-sectional drawing of a PVC joint.
(Courtesy NIBCO, Inc.)

Fig. 13-3). If the pipe will not enter the socket by that amount, the diameter may be reduced by sanding or filing. Extreme care should be taken not to gouge or flatten the pipe end when reducing the diameter. Unlike copper fittings, the inside walls of plastic fitting sockets are tapered so that the pipe makes contact with the sides of the fitting wall before the pipe reaches the seat of the socket.

Surfaces to be joined should be clean and free of moisture before application of the cement. The outside surface of the pipe (for socket depth) and the mating socket surface should be cleaned and the gloss removed with the recommended chemical cleaner.

An equally acceptable substitute is to remove the gloss from the mating surfaces (both pipe and socket) with abrasive paper or steel wool. Wipe off all particles of abrasive and/or PVC before applying primer or cement.

Apply Primer (PVC and CPVC Only)

Use only primer formulated for PVC or CPVC. Apply first to the inside of the fitting (see Fig. 13-4) and then to the outside of the pipe to the depth that will be taken into the fitting when seated. Wait 5 to 15 seconds before applying cement. A primer is not needed for ABS.

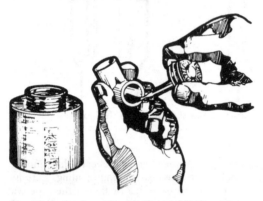

Fig. 13-4 Apply primer (PVC and CPVC only).
(Courtesy NIBCO, Inc.)

Apply Cement

When handling cement, keep the cement can closed and in a shady place when not in use. Discard the cement when an appreciable change in viscosity takes place, when a gel condition is indicated by the cement not flowing freely from the brush, or when cement appears lumpy and stringy. The cement should not be thinned. Keep the brush immersed in cement between applications.

Apply the solvent cement while the surfaces are still wet from the primer. Brush on a full, even coating to the inside of the fitting. Be careful not to form a puddle in the bottom of the fitting. Next, apply solvent to the pipe to the same depth as you applied primer. Recoat the pipe with a second uniform coat of cement, including the cut end of the pipe (see Fig. 13-5).

Fig. 13-5 Apply solvent cement. *(Courtesy NIBCO, Inc.)*

Cement is applied with a natural bristle or nylon brush. Use a $1/2$-inch (13-mm) brush for nominal pipe sizes $1/2$ inch (13 mm) and less. Use a 1-inch (24-mm) brush for pipe up through 2 inches (51 mm) nominal size and a brush width at least half the nominal pipe size for sizes above 2 inches (51 mm). For pipe sizes 6 inches (152 mm) and larger, a $2^1/_2$-inch (64-mm) brush is adequate.

Special Instructions for Bell End Pipe

The procedure as stated thus far may be followed in the case of bell end pipe, except that great care should be taken not to apply an excess of cement in the bell socket, and no cement should be applied on the bell-to-pipe transition area. This precaution is particularly important for installation of bell end pipe with a wall thickness of less than $1/8$ inch (3 mm).

Assembly of Joint

Solvent cement dries quickly. Work fast. Immediately after applying the last coat of cement to the pipe, insert the pipe into the fitting until it bottoms at the fitting shoulder (see Fig. 13-6). Turn the pipe or fitting a quarter-turn during assembly (but not after the pipe is bottomed) to evenly distribute the cement. Assembly should be completed within 20 seconds (longer in cold weather) after the last application of cement. The pipe should be inserted with a steady, even motion. Hammer blows should not be used.

Fig. 13-6 Fit and position pipe and fitting. *(Courtesy NIBCO, Inc.)*

Until the cement is set in the joint, the pipe may back out of the fitting socket if not held in place for approximately 1 minute after assembly.

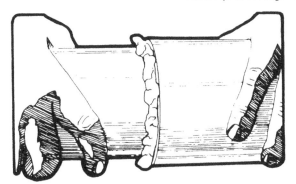

Fig. 13-7 Check for the correct bead. *(Courtesy NIBCO, Inc.)*

Care should be taken during assembly not to disturb or apply any force to joints previously made. Fresh joints can be destroyed by early rough handling.

A properly made joint will normally show a bead (see Fig. 13-7) around its entire perimeter. After assembly, wipe excess cement (see Fig. 13-8) from the pipe at the end of the fitting socket. Any gaps in the bead may indicate a defective assembly job due to insufficient cement or use of light-bodied cement on a gap fit where heavy-bodied cement should have been used.

Set Time

Handle the newly assembled joints carefully until the cement has gone through the set period. Recommended set time is related to temperature.

Plastic Pipe and Fittings

After the set period, the pipe can be carefully placed in position. If pipe is to be buried, shade backfill, leaving all joints exposed so that they can be examined during pressure tests.

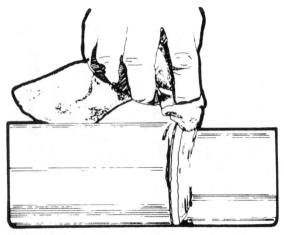

Fig. 13-8 Wipe off excess cement. *(Courtesy NIBCO, Inc.)*

Test pressure should be 150 percent of system design pressure and should be held until the system is checked for leaks, or you can follow the requirements of the applicable code, whichever is higher.

Table 13-2 shows water pressure ratings for PVC, and Table 13-3 shows ratings for ABS.

Table 13-1 Set Times

Temperature	Set Time
60°F to 100°F (15.5°C to 37.7°C)	30 minutes
40°F to 60°F (4.4°C to 15.5°C)	1 hour
20°F to 40°F (–6.6°C to 4.4°C)	2 hours
0°F to 20°F (–17.7°C to –6.6°C)	4 hours

Table 13-2 PVC Water Pressure Ratings at 73.4°F (23°C) for Schedule 40

No. PVC-1120-B PVC-1220-B Pipe PVC-2120-B Size CPVC-4120-B	PVC-2110-B		PVC-2112-B		PVC-2116-B		CPVC-4116-B	
	psi	kPa	psi	kPa	psi	kPa	psi	kPa
3/8 inch (9.5 mm)	620	4275	310	2137	390	2689	500	3447
1/2 inch (12.7 mm)	600	4137	300	2068	370	2551	480	3310
3/4 inch (19 mm)	480	3310	240	1655	300	2068	390	2689
1 inch (25.4 mm)	450	3103	220	1517	280	1931	360	2482
1 1/4 inches (31.75 mm)	370	2551	180	1241	230	1586	290	2000
1 1/2 inches (38 mm)	330	2275	170	1172	210	1448	260	1793
2 inches (51 mm)	280	1931	140	965	170	1172	220	1517
2 1/2 inches (63.5 mm)	300	2068	150	1034	190	1310	240	1655
3 inches (76 mm)	260	1793	130	896	160	1103	210	1448
3 1/2 inches (89 mm)	240	1655	120	827	150	1034	190	1310
4 inches (101.6 mm)	220	1517	110	758	140	965	180	1241
5 inches (127 mm)	190	1310	100	689	120	827	160	1103
6 inches (152.4 mm)	180	1241	90	621	110	758	140	965

Table 13-3 ABS Water Pressure Ratings at 73.4°F (23°C) for Schedule 40

Nominal Pipe Size	ABS-1210		ABS-1316		ABS-2112	
	psi	kPa	psi	kPa	psi	kPa
½ inch (12.7 mm)	298	2055	476	3282	372	2465
¾ inch (19 mm)	241	1662	385	2655	305	2103
1 inch (25.4 mm)	225	1551	360	2482	282	1944
1¼ inches (31.75 mm)	184	1269	294	2027	229	1579
1½ inches (38 mm)	165	1138	264	1820	207	1427
2 inches (51 mm)	139	958	222	1531	173	1193
2½ inches (63.5 mm)	152	1048	243	1675	190	1310
3 inches (76 mm)	132	910	211	1455	165	1138
4 inches (101.6 mm)	111	765	177	1220	138	951
6 inches (152.4 mm)	88	607	141	972	110	758

Note

For most cases, 48 hours is considered a safe period for the piping system to be allowed to stand vented to the atmosphere before pressure testing. Shorter periods may be satisfactory for high air temperatures, small sizes of pipe, quick-drying cement, and tight dry-fit joints.

PVC and ABS pipe and fittings may be stored either inside or outdoors if they are protected from direct sunlight. The plastic pipe should be stored in such a manner as to prevent sagging or bending. Plastic pipe should be supported in horizontal runs as shown in Table 13-4.

The industry does not recommend threading ABS or PVC Schedule 40 plastic pipe.

A quart can of the solvent recommended for the type of pipe being joined is generally sufficient for the average two-bath home.

Table 13-4 Horizontal Runs

Nominal Pipe Size	Schedule 40
1/2 inch and 3/4 inch (12.7 mm and 19 mm)	Every 4 ft. (1.22 m)
1 inch and 1 1/4 inches (25 mm and 32 mm)	Every 4 1/2 ft (1.37 m)
1 1/2 inches and 2 inches (38 mm and 51 mm)	Every 5 ft (1.52 m)
3 inches (76 mm)	Every 6 ft (1.83 m)
4 inches (102 mm)	Every 6 1/4 ft (1.9 m)
6 inches (152 mm)	Every 6 3/4 ft (2.06 m)
Nominal Pipe Size	**Schedule 80**
1/2 inch and 3/4 inch (12.7 mm and 19 mm)	Every 5 ft (1.52 m)
1 inch and 1 1/4 inches (25 mm and 32 mm)	Every 5 1/2 ft (1.68 m)
1 1/2 inches and 2 inches (38 mm and 51 mm)	Every 6 ft (1.83 m)
3 inches (76 mm)	Every 7 ft (2.13 m)
4 inches (102 mm)	Every 7 1/2 ft (2.29 m)
6 inches (152 mm)	Every 8 1/2 ft (2.59 m)

An ABS stack can be tested within 1 hour after the last joint is made up.

Common pipe dopes must not be used on threaded joints. Some pipe lubricants contain compounds that may soften the surface, which, under compression, can set up internal stress corrosion. If a lubricant is believed necessary, ordinary Vaseline or pipe tape can be used.

Do not use alcohol or antifreeze solutions containing alcohol to protect trap seals from freezing. Strong saline solutions or magnesium chloride in water (22 percent by weight) can be used safely. Glycerol (60 percent by weight) mixed with water is also recommended.

ABS absorbs heat so slowly that, once installed, heat from dishwashers and clothes washers and the discharge from similar installations will not cause any problem.

PVC solvent cements are available in two general viscosity categories: light and standard. The light cements are intended for use with pipes and fittings up through 2 inches (50.8 mm) N.P.S. and for pipes and fittings where interface fits occur between the parts to be joined.

Last, but not least, always look for the initials NSF. These initials stand for National Sanitation Foundation, denoting approval and standards met as handed down by NSF. The foundation is a nonprofit, noncommercial organization seeking solutions to all problems involving cleanliness. It is dedicated to the prevention of illness, the promotion of health, and the enrichment of the quality of American living through the improvement of the physical, biological, and social environment. The NSF seal on DWV and potable water plastic pipe and fittings indicates compliance with the foundation's policies and standards.

The NSF standards, research, and education programs are designed to benefit all parties: the manufacturer, the regulatory officials, the building industries, the product specifier, the installer, and ultimately the user. Their most important function may be their aid in providing protection to public health.

Solvent weld joints between plastic and metals are impossible for obvious reasons. There is no way to achieve a satisfactory bond. When making a transition between plastic and metal, temperature is a critical factor. For cold-water lines, you simply use a threaded adapter. However, this method is not recommended with hot water. Metal and plastic expand

and contract at different rates, working against each other in effect. Since a threaded adapter cannot compensate for this uneven expansion and contraction, the joint may eventually leak.

To solve this problem, a variety of transition adapters are available. However, some introduce serious flow restrictions. The CPVC-to-metal adapters shown in Fig. 13-9 were designed by NIBCO to avoid this problem.

Connecting CPVC to a Water Heater and Existing Plumbing

At least 12 inches (304.8 mm) of metal pipe is recommended on both hot- and cold-water lines above the tank (see Fig. 13-10) before converting to CPVC. Transition unions are used to connect the CPVC to the metal pipe risers.

Codes generally require the installation of a combination temperature/pressure relief valve on the heater tank. CPVC tubing may be used for the relief line to a point 6 inches (152.4 mm) above the floor. The relief line must be the same size as the outlet of the relief valve.

CPVC tubing should be supported at least every 3 feet (914.4 mm) in runs longer than 3 feet (914.4 mm). Some allowances must be made for expansion, especially on hot-water lines. A 12-inch (304.8-mm) offset every 10 feet (3.048 m) in a straight run is recommended.

Connecting ABS and PVC Drainage Systems to Other Materials

For DWV systems, most codes require that special adapter fittings be used when making a transition from one material to another. NIBCO offers a full selection of adapters, from threaded adapters (since there is very little thermal expansion in most DWV lines), to no-hub adapters for connecting to no-hub pipe.

198 Chapter 13

Copper end available in a variety of connections.

Buna-N Gasket for a tight seal.

CPVC Tailpiece solvent welds directly to tube.

Brass takeup nut tightens securely.

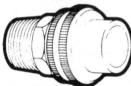
MPT to CPVC

FPT to CPVC

Fig. 13-9 NIBCO CPVC-to-metal unions and adapters.
(Courtesy NIBCO, Inc.)

Plastic Pipe and Fittings **199**

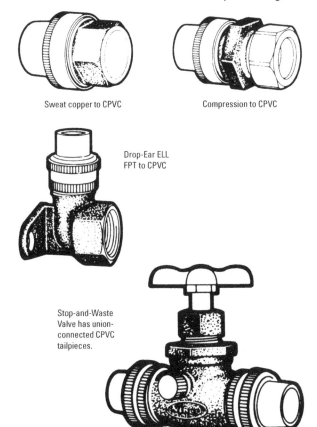

Sweat copper to CPVC

Compression to CPVC

Drop-Ear ELL
FPT to CPVC

Stop-and-Waste
Valve has union-
connected CPVC
tailpieces.

Fig. 13-9 *(continued)*

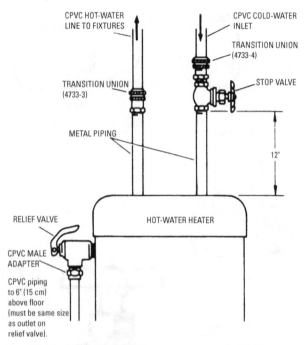

Fig. 13-10 Hot-water heater connections with CPVC.
(Courtesy NIBCO, Inc.)

Special lead-joint plastic adapters (see Fig. 13-11) for soil pipe are also offered. Plastic can be lead-caulked into cast iron without damage, since the lead cools much faster than plastic absorbs heat. However, the lead should not be heated beyond normal.

Plastic Pipe and Fittings **201**

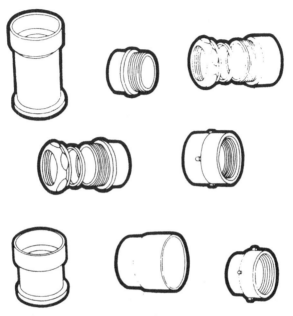

Fig. 13-11 Drainage system adapters. *(Courtesy NIBCO, Inc.)*

Note
Plastic pipe lighter than Schedule 80 should not be threaded. The wall thickness remaining after threading will not provide adequate strength.

Installing Multistory Stacks

Both ABS and PVC are used extensively in multistory installations. Although the thermal expansion and contraction rate of plastics is greater than that of metal, the *force* behind the

202 Chapter 13

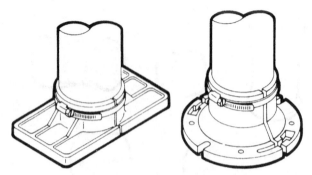

Fig. 13-12 Restraining fittings for multistory buildings.
(Courtesy NIBCO, Inc.)

expansion is not as great, so it is more easily controlled. For two-story buildings, there is no problem. For buildings that have three stories or more (model codes have approved plastic DWV for high-rise installations), use a restraining fitting every second floor (see Fig. 13-12). This fitting prevents vertical expansion (which could possibly damage the lines leading into the stack) but allows horizontal or circumferential expansion to relieve any stress.

Supporting Plastic DWV Systems

Plastic DWV systems should be supported, just like any other type of system. The vertical stack should be supported, but remember to allow for horizontal expansion. A noise problem is likely to arise if you block or wedge the stack against wood framing without space for expansion. Likewise, branch fittings serving trap arms should be supported where necessary, but remember to allow for expansion. When suspended below the floor, hanger support is recommended every 4–5 feet.

Allowing for Expansion/Contraction in a Plastic DWV System

Both ABS and PVC have a coefficient of expansion higher than most metal systems. Normal installations, with relatively short runs (less than 50 feet or 15.24 m), present little or no problem with proper support. The expansion and contraction will be contained between the supports, and the material will stress-relieve itself.

Where changes in direction are abrupt and followed by only a very short run (such as in a 45° or 90° offset), the support at the change of direction should be tightly clamped to the pipe. This avoids heavy flex loading of fittings at that point. Where changes in direction are followed by long, straight runs, supports next to the change in direction should be loose to allow movement of the pipe through the support.

Testing Time

The ABS stack can be tested 1 hour after the last joint is made. That means the warmer the ambient temperature, the faster the set.

PVC requires somewhat more time for initial set and curing.

Polybutylene Plumbing System

Polybutylene is said to offer a unique combination of flexibility, toughness, stress-crack resistance, and creep resistance. Fittings (see Fig. 13-13) are manufactured from Celcon, which is said to be impervious to scale buildup and corrosion. Certain fittings are also made from brass for special applications. Celcon is a widely accepted material for use in potable plumbing systems.

Polybutylene pipe resists corrosion, freezing, rust, acidic soils, and scale buildup. It will not burst when frozen and withstands temperatures of 180°F (82.2°C) at 100 psi (689 kPa). Polybutylene is also unaffected by freezing. The pipe

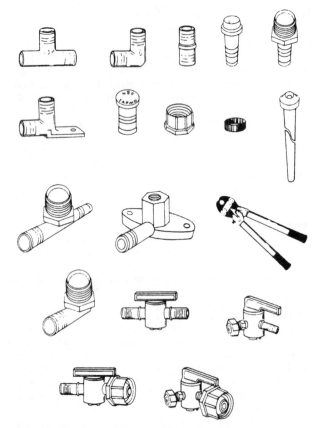

Fig. 13-13 Crimp fittings for PB plastic tubing.
(Courtesy Wrightway Mfg. Co.)

will expand to accommodate the ice. There is no splitting or cracking as with other materials.

It weighs less than 50 lbs (22.68 kg) for every 1000 feet (304.8 m) of ½-inch (12.7-mm) nominal tubing. Standard nominal sizes are ¼, ⅜, ½, ¾, 1, 1¼, 1½, and 2 inch (6.4, 9.5, 12.7, 19, 25, 32, 38, and 51 mm). Pipe up to 6 inches (15 cm) in diameter is available. Polybutylene is a poor conductor of heat, which means heat loss through pipes is reduced. With PB tubing, water hammer is virtually nonexistent, because PB is flexible.

Polybutylene pipe is recommended for these applications:

- In-house plumbing for hot- and cold-water systems
- Commercial water supply
- Slab heat (Most areas approve PB both above and in slab. Check your plumbing codes.)
- Hydronic heat
- Solar (Check with the manufacturer for specific applications.)

It is produced in rolls from 100 feet to 1000 feet (30.48 m to 304.8 m) and in 20-foot (6.1-m) straight lengths. It comes in three colors: white, gray, and black. Polybutylene should be stored out of direct sunlight because it is susceptible to ultraviolet attack. Black PB has the best ultraviolet protection.

Crimped Joints

Polybutylene plumbing systems are designed for the professional and, therefore, require special tools. The crimping tool must be calibrated at least twice a day. On the Wrightway CTGAGE, there are two steps, a go and no-go section, for each size of crimping tool. One gage works for all three crimp dimensions. The smaller diameter is called the go section. To use, simply place the gage inside the tool and crimp around the go section. Then, tighten or loosen the adjusting screw located near the top handle of the tool until the gage fits snugly. The gage should have some drag as it is turned inside the tool. Do not over-tighten.

A loose tool will result in an insufficient crimp, and a tool that is too tight may damage both the fitting and the tool. Remember to gage your tool before you start the day and during the day, depending on its use.

Note
> Lubricate the tool at least once a week to get maximum life. Check to make sure there is no looseness in the tool piece parts and that all assembly bolts are tight when adjusting the tool.

Following are step-by-step crimping instructions (see Fig. 13-14):

1. Cut the tubing squarely with a tubing cutter designed for plastic. The cutter should be rotated to slice into the tubing for a better cut.
2. Slip the crimp ring over the tubing.

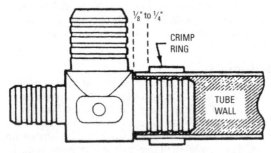

Fig. 13-14 Cross section of PB crimped joint. *(Courtesy U.S. Brass)*

3. Insert the fitting into the tube as far as possible. Position the ring over the center of the inserted fitting (approximately $1/4$ inch, or 6.4 mm, from the shoulder).
4. Set the crimping tool over the ring. Be sure the tool is square and centered on the ring. Close the tool until the stops are reached.
5. A properly crimped fitting should never leak or pull apart. A gage (QC43SP) to measure crimped connections is available from U.S. Brass. A $3/4$-inch (19.1-mm) connection should fit into the 0.960-inch (24.38-mm) end, and a $1/2$-inch (12.7-mm) connection should fit into the 0.715-inch (18.2-mm) end.
6. In the event of an improperly crimped fitting, remove the fitting and begin again at step 1.
7. On the threaded transition fittings, do not over-tighten. It is recommended that Teflon tape be used on all threaded connections.

Heat-Fusion Joints

The Wrightway heat-fusion tool shown in Fig. 13-15 must simply be plugged into any grounded 110-volt outlet. Approximately 15 minutes are needed for the tool to warm up. Care should be taken in securing the tool to avoid accidental burns.

The tool is designed with sockets and nipples that correspond to the tube size you are going to join. Take a piece of scrap tube and touch a socket or nipple. If the tube starts to soften and gives off a slight amount of smoke, your tool is ready for use. The smoke is nontoxic.

Place a brass ferrule inside of the tubing to be used. Its function is to keep the tube from collapsing during the fusion. Push a fitting onto the fusion tool nipple that corresponds to the size of the fitting. Place the tube with the ferrule inside

Fig. 13-15 Wrightway heat-fusion tool.
(Courtesy U.S. Brass)

the corresponding socket. Count to three. Pull the tube out of the socket, and then pull the fitting off the nipple. Push the tube into the socket until it stops. The joint is complete. The total operation should take less than 15 seconds.

What happened is that as you slid the tube into the fitting, you mixed the two melted surfaces, forming one piece. Because of the increased wall thickness (tube plus fitting), your joint is now stronger than the tube itself.

A few tips:

- Keep the tool clean. Use a fitting brush or steel wool.
- Do not overheat. You are overheating if the tube or fitting loses its shape.
- Do not join an overheated part. The tube will flare over the outside of the fitting.

Compression Fittings

An end-compression type of connection is shown in Fig. 13-16. It is a conventional-looking Celcon male fitting with threaded ends. The plumbing connection is made with a sealing cone, retaining ring, and nut placed on the tube. The

Fig. 13-16 Qicktite compression fitting. *(Courtesy U.S. Brass)*

nut-ring-cone assembly is simply threaded onto the fitting. *No special tools are required.* A pocketknife and pliers can do the job. The nut is hand-tightened and given a turn or two with pliers or a wrench. This connection is ideal for repairing plumbing and for nonprofessional use.

Qest threaded fittings use standard plumbing threads. They can be connected directly to *any other* standard plumbing thread. Nut-ring-cone Qicktite components can be applied directly to copper and CPVC tube, as well as to polybutylene.

Friction Welding

Friction welding is a method used by Qest in manufacturing supply tubes. The cone is spun onto the tube at 2500 rpm, creating friction and heat sufficient to weld the two pieces together. Portable friction-weld tools can be made available to professional plumbers and plumbing manufacturers.

14. CAST-IRON PIPE AND FITTINGS

As shown in Fig. 14-1, two separate systems constitute the home plumbing system: the water supply and the water disposal. Water in the *water supply system* is under about 50 pounds of pressure per square inch (psi). Water supply pipes can be somewhat small in diameter but still carry enough water. Water in the *disposal system* (also called the *drain-waste-vent*, or *DWV*, system) flows by gravity. Therefore, piping and fittings in the DWV system are larger in diameter to carry the required flow without clogging or backing up. Both systems operate safely if designed properly. For years, cast iron has been the material of choice for these two types of systems not only in homes but also in industry and business.

Organized in 1949 by the leading American manufacturers of cast-iron soil pipe and fittings, the Cast Iron Soil Pipe Institute (CISPI) sponsors a continuous program of product testing, evaluation, and development to improve the industry's products, while achieving standardization of cast-iron soil pipe and fittings. Members of the Cast Iron Soil Pipe Institute adopted an insignia for use as a symbol of quality as a means of providing a simple designation of desired standards of quality. At the present, the CISPI membership consists of the following 12 companies (operating 17 plants in 9 states):

- The American Brass & Iron Foundry
- Charlotte Pipe & Foundry Company
- Tyler Pipe Industries

This chapter examines fittings and specifications for the following two categories:

- Cast-iron soil pipe
- Cast-iron no-hub

212 Chapter 14

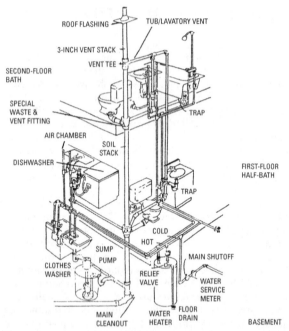

Fig. 14-1 Note the use of cast-iron pipe in the typical complete home plumbing system.

Cast-Iron Soil Pipe Fittings and Specifications

This section provides an examination of fittings and specifications of cast-iron soil pipe. This section pairs figures and specification tables for the following topic areas:

- $1/4$-bend fitting
- Low-hub $1/4$ bends
- Sweeps
- $1/6$, $1/8$, and $1/16$ bends
- Single and double sanitary T-branches
- Single and double Y-branches
- Single and double combination Y and $1/4$ bends
- Single and double T-branches
- Upright Y-branches
- $1/2$ S- or P-traps
- Reducers
- Vent branches
- $1/8$-bend offsets
- Minimum offsets for $1/6$, $1/8$, and $1/16$ bends

$1/4$ Bends

Fig. 14-2 shows a $1/4$-bend fitting, and Table 14-1 shows $1/4$-bend specifications.

Long Low-Hub $1/4$ Bends

Fig. 14-3 shows low-hub $1/4$ bends, while Table 14-2 shows long low-hub $1/4$ bends in inches and Table 14-3 shows long low-hub $1/4$ bends in metric.

Note
Where space is limited in reference to water closets, $1 3/4$ inches (44 mm) is gained in measurement C.

Sweeps

Fig. 14-4 shows sweeps, while Table 14-4 shows short-sweep specifications, Table 14-5 shows long-sweep specifications, and Table 14-6 shows reducing long-sweep specifications.

214 Chapter 14

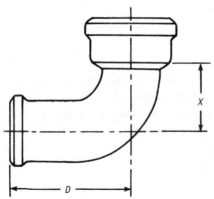

Fig. 14-2 ¼ bend.

Table 14-1 ¼ Bend

Size		D		X	
Inches	mm	Inches	mm	Inches	mm
2	51	6	152	3¼	83
3	76	7	178	4	102
4	102	8	203	4½	114
5	127	8½	216	5	127
6	152	9	229	5½	140
8	203	11½	292	6⅝	168
10	254	12½	318	7⅝	194

Cast-Iron Pipe and Fittings 215

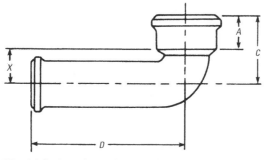

Fig. 14-3 Long low-hub ¼ bends.

Table 14-2 Long Low-Hub ¼ Bends (Inches)

Size	D	A	X	C
4 inches × 12 inches	12 inches	3 inches	2¾ inches	5¾ inches
4 inches × 14 inches	14 inches	3 inches	2¾ inches	5¾ inches
4 inches × 16 inches	16 inches	3 inches	2¾ inches	5¾ inches
4 inches × 18 inches	18 inches	3 inches	2¾ inches	5¾ inches

Table 14-3 Long Low-Hub ¼ Bends (Metric)

Size	D	A	X	C
102 × 305 mm	305 mm	76 mm	70 mm	146 mm
102 × 356 mm	356 mm	76 mm	70 mm	146 mm
102 × 406 mm	406 mm	76 mm	70 mm	146 mm
102 × 457 mm	457 mm	76 mm	70 mm	146 mm

216 Chapter 14

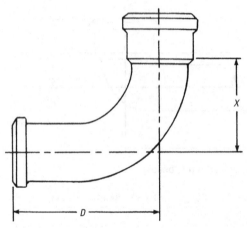

Fig. 14-4 Sweeps.

Table 14-4 Short Sweeps (Metric and English)

Size		D		X	
Inches	mm	Inches	mm	Inches	mm
2	51	8	203	5¼	133
3	76	9	229	6	152
4	102	10	254	6½	165
5	127	10½	267	7	178
6	152	11	279	7½	191
8	203	13½	343	8⅝	219
10	254	14½	368	9⅝	244

Cast-Iron Pipe and Fittings 217

Table 14-5 Long Sweeps

Size		D		X	
Inches	mm	Inches	mm	Inches	mm
2	51	11	279	$8^1/_4$	210
3	76	12	305	9	229
4	102	13	330	$9^1/_2$	241
5	127	$13^1/_2$	343	10	254
6	152	14	356	$10^1/_2$	267
8	203	$16^1/_2$	419	$11^5/_8$	295
10	254	$17^1/_2$	445	$12^5/_8$	321

Table 14-6 Reducing Long Sweeps

	Size	D	X
Inches	3 × 2	9	6
	4 × 3	10	$6^1/_2$
mm	76 × 51	229	152
	102 × 76	254	165

$^1/_6$, $^1/_8$, and $^1/_{16}$ Bends

Fig. 14-5 shows bends for $^1/_6$, $^1/_8$, and $^1/_{16}$, while Table 14-7 shows specifications for $^1/_6$ bends, Table 14-8 shows specifications for $^1/_8$ bends, and Table 14-9 shows specifications for $^1/_{16}$ bends.

Single and Double Sanitary T-Branches

Fig. 14-6 shows single and double sanitary T-branches, while Table 14-10 shows specifications for single and double sanitary T-branches in inches and Table 14-11 shows specifications for single and double sanitary T-branches in millimeters.

218 Chapter 14

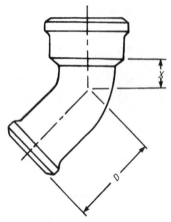

Fig. 14-5 1/6, 1/8, 1/16 bends.

Table 14-7 1/6 Bends

Size		D		X	
Inches	mm	Inches	mm	Inches	mm
2	51	4³/₄	121	2	51
3	76	5¹/₂	140	2¹/₂	64
4	102	6⁵/₁₆	160	2¹³/₁₆	71
5	127	6⁵/₈	168	3¹/₈	79
6	152	6⁷/₈	175	3³/₈	86
8	203	9	229	4¹/₈	105
10	254	9⁹/₁₆	243	4¹¹/₁₆	119

Cast-Iron Pipe and Fittings **219**

Table 14-8 $^1/_8$ Bends

Size		D		X	
Inches	mm	Inches	mm	Inches	mm
2	51	$4^1/_4$	108	$1^1/_2$	38
3	76	$4^{15}/_{16}$	125	$1^{15}/_{16}$	49
4	102	$5^{11}/_{16}$	144	$2^3/_{15}$	56
5	127	$5^7/_8$	149	$2^3/_5$	60
6	152	$6^1/_{16}$	154	$2^9/_{16}$	65
8	203	8	203	$3^1/_{16}$	79
10	254	$8^3/_8$	213	$3^1/_2$	89

Table 14-9 $^1/_{16}$ Bends

Size		D		X	
Inches	mm	Inches	mm	Inches	mm
2	51	$3^5/_8$	92	$^7/_8$	22
3	76	$4^3/_{16}$	106	$1^3/_{16}$	30
4	102	$4^{13}/_{16}$	122	$1^5/_{16}$	33
5	127	$4^7/_8$	124	$1^3/_8$	35
6	152	5	127	$1^1/_2$	38
8	203	$6^{11}/_{16}$	170	$1^{13}/_{16}$	46
10	254	$6^7/_8$	175	2	51

220 Chapter 14

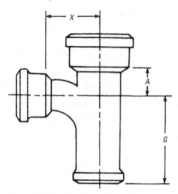

Fig. 14-6 Single and double sanitary T-branches.

Table 14-10 Single and Double Sanitary T-Branches (Inches)

Size	X	G	A
2	2³/₄	6¹/₄	1³/₄
3	4	7¹/₂	2¹/₂
4	4¹/₂	8	3
5	5	8¹/₂	3¹/₂
6	5¹/₂	9	4
3 × 2	4	7	2
4 × 2	4¹/₂	7	2
4 × 3	4¹/₂	7¹/₂	2¹/₂
5 × 2	5	7	2
5 × 3	5	7¹/₂	2¹/₂
5 × 4	5	8	3
6 × 2	5¹/₂	7	2
6 × 3	5¹/₂	7¹/₂	2¹/₂
6 × 4	5¹/₂	8	3
6 × 5	5¹/₂	8¹/₂	3¹/₂

Table 14-11 Single and Double Sanitary T-Branches (mm)

Size	X	G	A
51	70	159	44
76	102	191	64
102	114	203	76
127	127	216	89
152	140	229	102
76 × 51	102	178	51
102 × 51	114	178	51
102 × 76	114	191	64
127 × 51	127	178	51
127 × 76	127	191	64
127 × 102	127	203	76
152 × 51	140	178	51
152 × 76	140	191	64
152 × 102	140	203	76
152 × 127	140	216	89

Single and Double Y-Branches

Fig. 14-7 shows single and double Y-branches, while Table 14-12 shows specifications for single and double Y-branches in inches and Table 14-13 shows the specifications in millimeters.

Single and Double Combination Y and $^1/_4$ Bends

Fig. 14-8 shows single and double combination Y and $^1/_4$ bends, while Table 14-14 shows specifications for single and double combination Y and $^1/_4$ bends in inches and Table 14-15 shows specifications in millimeters.

222 Chapter 14

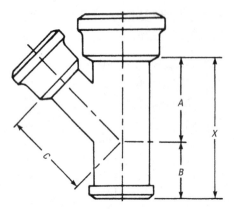

Fig. 14-7 Y-branches, single and double.

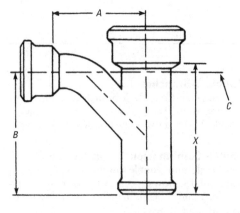

Fig. 14-8 Single and double combination Y and 1/4 bends.

Cast-Iron Pipe and Fittings **223**

Table 14-12 Y-Branches, Single and Double (Inches)

Size	A	B	C	X
2	4	4	4	8
3	5½	5	5½	10½
4	6¾	5¼	6¾	12
5	8	5½	8	13½
6	9¼	5¾	9¼	15
8	11¹³/₁₆	7¹¹/₁₆	11¹³/₁₆	19½
10	14½	8	14½	22½
3 × 2	4¹³/₁₆	4³/₁₆	5	9
4 × 2	5⅜	3⅝	5¾	9
4 × 3	6¹/₁₆	4⁷/₁₆	6¼	10½
5 × 2	5⅞	3⅛	6½	9
5 × 3	6⅝	3⅞	7	10½
5 × 4	7⁵/₁₆	4¹¹/₁₆	7½	12
6 × 2	6⁷/₁₆	2⁹/₁₆	7¼	9
6 × 3	7⅛	3⅜	7¾	10½
6 × 4	7¹³/₁₆	4¹³/₁₆	8¼	12
6 × 5	8⁹/₁₆	4¹⁵/₁₆	8¾	13½

Table 14-13 Y-Branches, Single and Double (mm)

Size	A	B	C	X
51	102	102	102	203
76	140	127	140	267
102	171	133	171	305
127	203	140	203	343
152	235	146	235	381
203	300	195	300	495
254	368	203	368	572
76 × 51	122	106	127	229
102 × 51	137	92	146	229
102 × 76	154	113	159	267
127 × 51	149	79	165	229
127 × 76	168	98	178	267
127 × 102	186	119	191	305
152 × 51	164	65	184	229
152 × 76	181	86	197	267
152 × 102	198	106	210	305
152 × 127	217	125	222	343

Table 14-14 Single and Double Combination Y and $\frac{1}{4}$ Bends (Inches)

Size	A	B	X
2	$4^7/_8$	$7^3/_8$	8
3	7	$10^1/_{16}$	$10^1/_2$
4	9	12	12
5	11	$14^1/_3$	$13^1/_2$
6	$12^7/_8$	$16^1/_{16}$	15
8	17	$21^9/_{16}$	$19^1/_2$
3 × 2	$5^3/_4$	$8^3/_{16}$	9
4 × 2	$6^1/_4$	$8^3/_{16}$	9
4 × 3	$7^1/_2$	$10^1/_{16}$	$10^1/_2$
5 × 2	$6^3/_4$	$8^3/_8$	9
5 × 3	8	$10^1/_{16}$	$10^1/_2$
5 × 4	$9^1/_2$	12	12
6 × 2	$7^1/_4$	$8^3/_{16}$	9
6 × 3	$8^1/_2$	$10^1/_{16}$	$10^1/_2$
6 × 4	10	12	12
6 × 5	$11^1/_2$	$14^3/_{16}$	$13^1/_2$

226 Chapter 14

Table 14-15 Single and Double Combination Y and
$^{1}/_{4}$ Bends (mm)

Size	A	B	X
51	124	187	203
76	178	256	267
102	229	305	305
127	279	359	343
152	327	408	381
203	432	548	495
76 × 51	146	208	229
102 × 51	159	208	229
102 × 76	191	256	267
127 × 51	171	213	229
127 × 76	203	256	267
127 × 102	241	305	305
152 × 51	184	208	229
152 × 76	216	256	267
152 × 102	254	305	305
152 × 127	292	360	343

Single and Double T-Branches
Fig. 14-9 shows single and double T-branches, while Table 14-16 shows specifications for single and double T-branches in inches and Table 14-17 shows specifications in millimeters.

Upright Y-Branches
Fig. 14-10 shows upright Y-branches, and Table 14-18 shows specifications for upright Y-branches in inches and Table 14-19 shows specifications in millimeters.

Cast-Iron Pipe and Fittings **227**

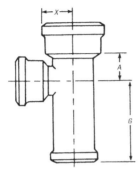

Fig. 14-9 Single and double T-branches.

Table 14-16 Single and Double T Branches (Inches)

Size	X	G	A
2	1¾	6¼	1¾
3	2½	7½	2½
4	3	8	3
5	3½	8½	3½
6	4	9	4
3 × 2	2½	7	2
4 × 2	3	7½	2
4 × 3	3	7	2½
5 × 2	3½	7	2
5 × 3	3½	7½	2½
5 × 4	3½	8	3
6 × 2	4	7	2
6 × 3	4	7½	2½
6 × 4	4	8	3
6 × 5	4	8½	3½

Table 14-17 Single and Double T-Branches (mm)

Size	X	G	A
51	44	159	44
76	64	191	64
102	76	203	76
127	89	216	89
152	102	229	102
76 × 51	64	178	51
102 × 51	76	178	51
102 × 76	76	191	64
127 × 51	89	178	51
127 × 76	89	191	64
127 × 102	89	203	76
152 × 51	102	178	51
152 × 76	102	191	64
152 × 102	102	203	76
152 × 127	102	216	89

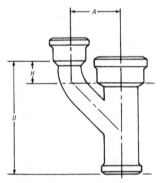

Fig. 14-10 Upright Y-branches.

Cast-Iron Pipe and Fittings **229**

Table 14-18 Upright Y-Branches (Inches)

Size	A	H	X	D
2	$4\frac{1}{2}$	2	8	10
3	$5\frac{1}{2}$	$1^{15}/_{16}$	$10\frac{1}{2}$	$12^{7}/_{16}$
4	$6\frac{1}{2}$	$1^{15}/_{16}$	12	$13^{15}/_{16}$
3 × 2	5	$1^{15}/_{16}$	9	$10^{15}/_{16}$
4 × 2	$5\frac{1}{2}$	$1^{15}/_{16}$	9	$10^{15}/_{16}$
4 × 3	6	$1^{15}/_{16}$	$10\frac{1}{2}$	$12^{7}/_{16}$

Table 14-19 Upright Y-Branches (mm)

Size	A	H	X	D
51	114	51	203	254
76	140	49	267	316
102	165	49	305	354
76 × 51	127	49	229	278
102 × 51	140	49	229	278
102 × 76	152	49	267	316

Cleanout T-Branches

Fig. 14-11 shows a cleanout T-branch, and Table 14-20 shows specifications for cleanout T-branches.

$\frac{1}{2}$ S- or P-Traps

Fig. 14-12 shows $\frac{1}{2}$ S- or P-traps, and Table 14-21 shows specifications for $\frac{1}{2}$ S- or P-traps.

Reducers

Fig. 14-13 shows a reducer, and Table 14-22 shows specifications for reducers.

230 Chapter 14

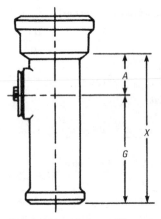

Fig. 14-11 Cleanout T-branch.

Table 14-20 Cleanout T-Branches

Size		A		G		X	
Inches	mm	Inches	mm	Inches	mm	Inches	mm
2	51	1³/₄	44	6¹/₄	159	8	203
3	76	2¹/₂	64	7¹/₂	191	10	254
4	102	3	76	8	203	11	279
5	127	3¹/₂	89	8¹/₂	216	12	305
6	152	4	102	9	229	13	330

Cast-Iron Pipe and Fittings **231**

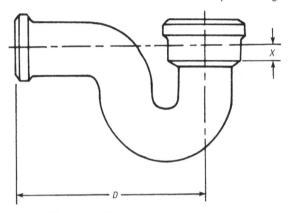

Fig. 14-12 $\frac{1}{2}$ S- or P-traps.

Table 14-21 $\frac{1}{2}$ S- or P-Traps

Size		D		X	
Inches	mm	Inches	mm	Inches	mm
2	51	$9\frac{1}{2}$	241	$1\frac{1}{2}$	38
3	76	12	305	$1\frac{1}{4}$	32
4	102	14	356	1	25
5	127	$15\frac{1}{2}$	394	$\frac{1}{2}$	13
6	152	17	432	0	0

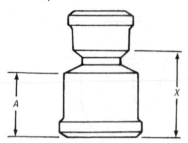

Fig. 14-13 Reducers.

Table 14-22 Reducers

Size		A		X	
Inches	mm	Inches	mm	Inches	mm
3 × 2	76 × 51	3¼	83	4¾	121
4 × 2	102 × 51	4	102	5	127
4 × 3	102 × 76	4	102	5	127
5 × 2	127 × 51	4	102	5	127
5 × 3	127 × 76	4	102	5	127
5 × 4	127 × 102	4	102	5	127
6 × 2	152 × 51	4	102	5	127
6 × 3	152 × 76	4	102	5	127
6 × 4	152 × 102	4	102	5	127
6 × 5	152 × 127	4	102	5	127

Vent Branches

Fig. 14-14 shows a vent branch, and Table 14-23 shows specifications for vent branches.

Cast-Iron Pipe and Fittings 233

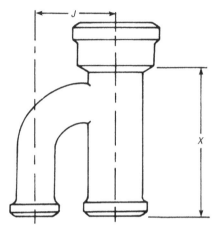

Fig. 14-14 Vent branches.

Table 14-23 Vent Branches

Size		J		X	
Inches	mm	Inches	mm	Inches	mm
2	51	$4\frac{1}{2}$	114	8	203
3	76	$5\frac{1}{2}$	140	10	254
4	102	$6\frac{1}{2}$	165	11	279
3×2	76×51	5	127	9	229
4×2	102×51	$5\frac{1}{2}$	140	9	229
4×3	102×76	6	152	10	254

$1/_8$-Bend Offsets

Fig. 14-15 shows a $1/_8$-bend offset, while Table 14-24 shows specifications for $1/_8$-bend offsets in inches and Table 14-25 shows specifications in millimeters.

Minimum Offsets Using $1/_6$-, $1/_8$-, $1/_{16}$-Bend Fittings

Fig. 14-16 shows minimum offsets for $1/_6$, $1/_8$, and $1/_{16}$ bends, while Table 14-26 shows specifications for minimum offsets for $1/_6$-bend fittings, Table 14-27 shows specifications for minimum offsets for $1/_8$-bend fittings, and Table 14-28 shows specifications for minimum offsets for $1/_{16}$-bend fittings.

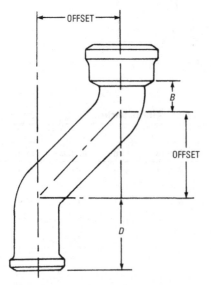

Fig. 14-15 $1/_8$-bend offset.

Table 14-24 $1/_8$-Bend Offset (Inches)

Size	Offset	Hub	D	B
2	(2–4–6–8–10)	$2^1/_2$	$4^1/_2$	1
3	(2–4–6–8–10)	$2^3/_4$	5	$1^1/_2$
4	(2–4–6–8–10)	3	$5^1/_4$	$1^3/_4$
5	(2–4–6–8–10)	3	$5^9/_{16}$	$1^{13}/_{16}$
6	(2)	3	$5^5/_8$	2
6	(4–6–8–10)	3	$5^{13}/_{16}$	$2^3/_{16}$

Note: $1/_8$ -bend offset fittings made in pipe sizes 2 inches through 6 inches (offsets 2 inches through 18 inches).

Table 14-25 $1/_8$-Bend Offset (mm)

Size	Offset	Hub	D	B
51	(51–102–152–203–254)	64	108	25
76	(51–102–152–203–254)	70	127	38
102	(51–102–152–203–254)	76	133	44
127	(51–102–152–203–254)	76	141	49
152	(51)	76	143	51
152	(102–152–203–254)	76	149	56

Note: $1/_8$ -bend offset fittings made in pipe sizes 51 mm through 152 mm (offsets 51 mm through 457 mm).

Cast-Iron No-Hub Pipe Fittings and Specifications

The most outstanding advantages of cast-iron (CI) no-hub joints are as follows:

- Faster installation
- More economical

236 Chapter 14

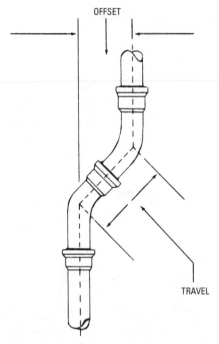

Fig. 14-16 Minimum offsets using 1/6-, 1/8-, 1/16-bend fittings.

- Space-saving (3-inch, or 76-mm, size fits neatly in 2-inch × 4-inch, or 50.8-mm × 101.6-mm, framed wall)
- No waste
- Quieter
- No vibration

Table 14-26 Minimum Offsets Using $^1/_6$-Bend Fittings

Size		Travel		Minimum Offset	
Inches	mm	Inches	mm	Inches	mm
2	51	$6^3/_4$	171	$5^7/_8$	149
3	76	8	203	$6^{15}/_{16}$	176
4	102	$9^1/_8$	232	$7^{15}/_{16}$	202
5	127	$9^3/_4$	248	$8^1/_2$	216
6	152	$10^1/_4$	260	$8^7/_8$	225
8	203	$13^1/_8$	333	$11^3/_8$	289
10	254	$14^1/_4$	362	$12^3/_8$	314
12	305	17	432	$14^3/_4$	375

Table 14-27 Minimum Offsets Using $^1/_8$-Bend Fittings

Size		Travel		Minimum Offset	
Inches	mm	Inches	mm	Inches	mm
2	51	$5^3/_4$	146	$4^3/_{32}$	104
3	76	$6^7/_8$	175	$4^7/_8$	124
4	102	$7^7/_8$	200	$5^9/_{16}$	141
5	127	$8^1/_4$	210	$5^7/_8$	149
6	152	$8^5/_8$	219	$6^1/_8$	156
8	203	$11^1/_8$	283	$7^7/_8$	200
10	254	$11^7/_8$	302	$8^7/_{16}$	214
12	305	$14^3/_8$	365	$10^1/_4$	260

- Lightweight
- Physically less taxing than conventional jointing
- Testing takes less time (five floors can be tested at one time)

Table 14-28 Minimum Offsets Using $^1/_{16}$-Bend Fittings

Size		Travel		Minimum Offset	
Inches	mm	Inches	mm	Inches	mm
2	51	$4^1/_2$	114	$1^3/_4$	44
3	76	$5^3/_8$	137	$2^1/_{16}$	52
4	102	$6^1/_8$	156	$2^5/_{16}$	59
5	127	$6^1/_4$	159	$2^3/_8$	60
6	152	$6^1/_2$	165	$2^7/_{16}$	62
8	203	$8^1/_2$	216	$3^1/_4$	83
10	254	$8^7/_8$	225	$3^3/_8$	86
12	305	11	279	$4^1/_4$	108

Overview

These suggestions are for use with the CI no-hub system utilizing a neoprene sleeve-type coupling device consisting of an internally ribbed *elastomeric sealing sleeve* within a protective corrugated stainless steel shield band secured by two stainless steel bands with tightening devices (also of stainless steel).

During installation assembly, CI no-hub pipe and fittings must be inserted into the sleeve and firmly seated against the center rib or shoulder of the gasket. To provide a sound joint with field-cut lengths of pipe, it is necessary to have ends cut smooth and as square as possible. Snap- or abrasive-type cutters may be used.

The stainless steel bands must be tightened alternately and firmly to not less than 48 (nor more than 60) inch-lbs (67.79 N-m) of torque.

Installation

Figs. 14-17 through 14-20 show proper installation techniques for cast-iron no-hub pipe fittings.

Cast-Iron Pipe and Fittings 239

Fig. 14-17 No-hub cast-iron pipe ready for joining. Note extreme simplicity of joint parts.

Fig. 14-18 Sleeve coupling is placed on one end of pipe. Stainless steel shield and band clamps are placed on the end of the other pipe. Two bands are used on pipe sizes $1\frac{1}{2}$ inches to 4 inches; four bands are used on sizes 5 inches through 10 inches.

240 Chapter 14

Fig. 14-19 Pipe ends are butted against integrally molded shoulder inside of the sleeve. Shield is slid into position and tightened to make a joint that is quickly assembled and permanently fastened.

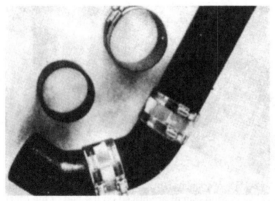

Fig. 14-20 Fittings are jointed and fastened in the same way. Making up joints in close quarters is easily done.

Vertical Piping

Secure vertical piping at sufficiently close intervals to keep the pipe in alignment and to support the weight of the pipe and its contents. Support stacks at their bases and at each floor (see Fig. 14-21).

Horizontal Piping Suspended

Support ordinary horizontal piping and fittings at sufficiently close intervals to maintain alignment and to prevent sagging or grade reversal.

Support each length of pipe by a hanger located as near the coupling as possible, but not more than 18 inches (457.2 mm) from the joint (see Fig. 14-22).

If piping is supported by nonrigid hangers more than 18 inches (457.2 mm) long, install sufficient sway bracing to prevent lateral movement, such as might be caused by seismic shock (see Fig. 14-23).

Hangers should also be provided at each horizontal branch connection (see Fig. 14-24).

Horizontal Piping Underground

CI no-hub systems laid in trenches should be continuously supported on undisturbed earth, on compacted fill of selected material, or on masonry blocks at each coupling.

To maintain proper alignment during backfilling, stabilize the pipe in the proper position by partial backfilling and cradling, or by the use of adequate metal stakes or braces fastened to the pipe.

Piping laid on grade should be adequately staked to prevent misalignment when the slab is poured.

Vertical sections and their connection branches should be adequately staked and fastened to driven pipe or reinforcing rod to keep them stable while backfill is placed or concrete is poured.

Figs. 14-25 through 14-34 show techniques for horizontal piping underground.

242 Chapter 14

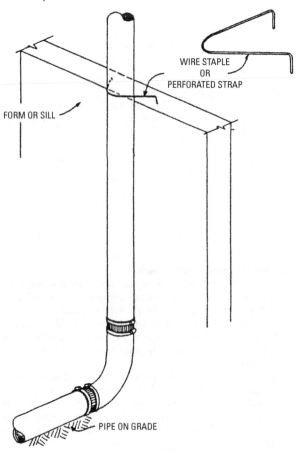

Fig. 14-21 Support for vertical pipe.

Cast-Iron Pipe and Fittings **243**

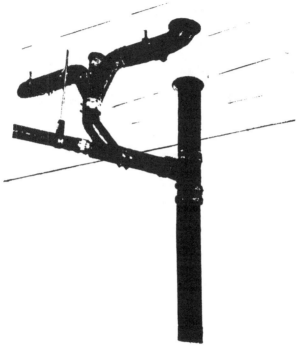

Fig. 14-22 Extra support is needed with no-hub. Note conventional hangers and supports in the ceiling to make the system more rigid.

Fig. 14-23 Sway brace and method of hanging.

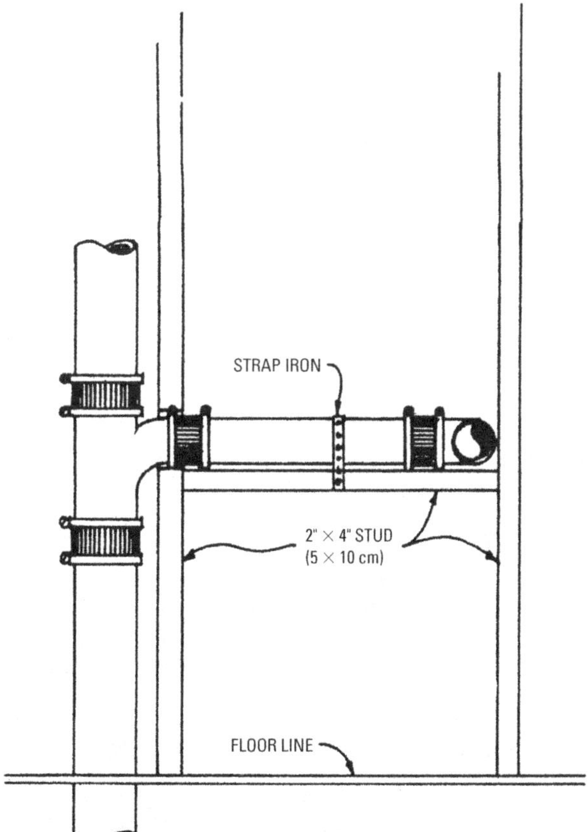

Cast-Iron Pipe and Fittings **245**

Fig. 14-24 Strapping horizontal run to a cross brace.

246 Chapter 14

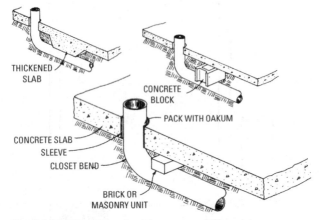

Fig. 14-25 Pipe through various types of floor slabs.

Fig. 14-26 Note sway brace and the method of hanging and cleanout.

Cast-Iron Pipe and Fittings **247**

Fig. 14-27 Note the neat method of hanging pipe.

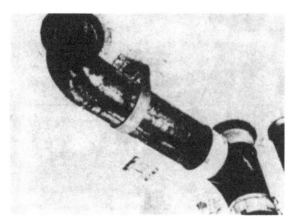

Fig. 14-28 Method of using hanger for a closet bend. Note sleeves and oakum in sleeves.

248 Chapter 14

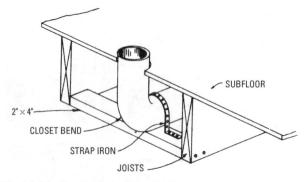

Fig. 14-29 Bracing for a closet bend.

Fig. 14-30 View showing method of hanging and sway bracing.

Cast-Iron Pipe and Fittings 249

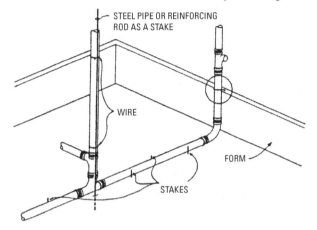

Fig. 14-31 Slab-on-grade installation.

Fig. 14-32 Method of clamping the no-hub pipe at each floor, using a friction clamp or floor clamp.

Cast-Iron Pipe and Fittings 251

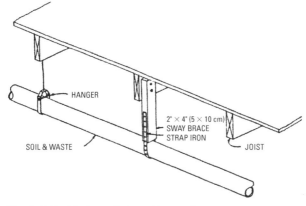

Fig. 14-33 Horizontal pipe with sway brace.

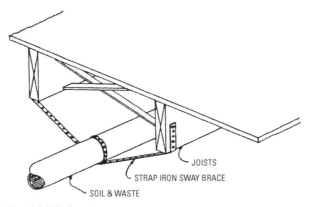

Fig. 14-34 Sway brace.

Remember
The spacer inside of the neoprene gasket where fittings or pipe ends meet measures $3/32$ inch (2.38 mm).

Note
Fitting measurements, laying lengths, may vary $1/8$ inch (3.2 mm) plus or minus. Five-foot (152-cm) lengths of pipe may vary $1/4$ inch (6.4 mm) plus or minus; 10-inch (305-cm) lengths of pipe may vary $1/2$ inch (13 mm) plus or minus. All 2-inch, 3-inch, and 4-inch (51-, 76-, and 102-mm) fittings, and so on, lay the same length; this is an advantage in that one fitting can be removed from a line and another inserted without cutting the pipe. No-hub pipe may be cast with or without a spigot bead and positioning lugs.

Specifications
The remainder of this chapter provides specifications for various types of cast-iron no-hub pipe fittings in the following topic areas:

- Single and double Y-branches
- Single and double combination Y and $1/4$ bend
- $1/5$, $1/6$, $1/8$, $1/16$ bends
- Sweeps
- $1/4$ bends
- Test tees
- Single and double sanitary T-branches
- Single and double sanitary T-branches, tapped
- Y-branches, tapped
- Upright Y-branches
- $1/4$ bend, double
- $1/4$ bend, short-radius, tapped

Cast-Iron Pipe and Fittings 253

- Spread between vent and revent when using a wye and ¼ bend
- ½ S- or P-traps
- Tapped extension place
- Increaser-reducer
- Minimum offsets using bend fittings

Single and Double Y-Branches

Fig. 14-35 shows single and double Y-branches, while Table 14-29 shows specifications for single and double Y-branches in inches and Table 14-30 shows specifications in millimeters.

Single and Double Combination Y and ¼ Bends

Fig. 14-36 shows single and double combination Y and ¼ bends, while Table 14-31 shows specifications for single

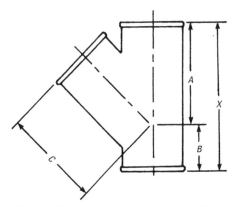

Fig. 14-35 Y-branches, single and double.

Table 14-29 Y-Branches, Single and Double (Inches)

Size	A	B	C	X
1½	4	2	4	6
2	4⅝	2	4⅝	6⅝
3	5¾	2¼	5¾	8
4	7 1/16	2 7/16	7 1/16	9½
5	9½	3⅛	9½	12⅝
6	10¾	3 5/16	10¾	14 1/16
3 × 2	5⅛	1½	5 5/16	6⅝
4 × 2	5⅝	1	6	6⅝
4 × 3	6 5/16	1 11/16	6½	8
5 × 2	7⅛	15/16	7½	8 1/16
5 × 3	8	1 11/16	8	9 11/16
5 × 4	8¾	2 7/16	8⅛	11 3/16
6 × 2	7 13/16	½	8¼	8 5/16
6 × 3	8½	1¼	8¾	9¾
6 × 4	9¼	1 15/16	9¼	11 3/16
6 × 5	9 15/16	2 9/16	10¼	12½

and double combination Y and ¼ bends in inches and Table 14-32 shows specifications in millimeters.

⅕, ⅙, ⅛, 1/16 Bends

Fig. 14-37 shows bends for ⅕, ⅙ ⅛, and 1/16, while Table 14-33 shows specifications for ⅕ bends, Table 14-34 shows specifications for ⅙ bends, Table 14-35 shows specifications for ⅛ bends, and Table 14-36 shows specifications for 1/16 bends.

Cast-Iron Pipe and Fittings **255**

Table 14-30 **Y-Branches, Single and Double (mm)**

Size	A	B	C	X
38	102	51	102	152
51	117	51	117	168
76	146	57	146	203
102	179	62	179	241
127	241	79	241	321
152	273	84	273	357
76 × 51	130	38	135	168
102 × 51	143	25	152	168
102 × 76	160	43	165	203
127 × 51	181	24	191	205
127 × 76	203	43	203	246
127 × 102	222	62	216	284
152 × 51	198	13	210	211
152 × 76	216	32	222	248
152 × 102	235	49	235	284
152 × 127	252	64	260	318

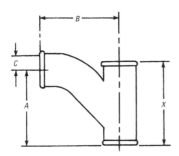

Fig. 14-36 Single and double combination Y and $1/4$ bend.

Table 14-31 Single and Double Combination Y and ¼ Bends (Inches)

Size	A	B	C	X
1½	4¾	5⅜	1¼	6
2	5⅜	6⅛	1¼	6⅝
3	7⁵⁄₁₆	8	¹¹⁄₁₆	8
4	9¼	10	¼	9½
5	11¾	12½	⅞	12⅝
6	13⅝	14⅜	⁷⁄₁₆	14¹⁄₁₆
2 × 1½	5	5⅝	1	6
3 × 2	5½	6¾	1⅛	6⅝
4 × 2	5½	7¼	1⅛	6⅝
4 × 3	7¼	8½	¾	8
5 × 2	5¹⁵⁄₁₆	7¾	2⅛	8¹⁄₁₆
5 × 3	7¾	9	1¹⁵⁄₁₆	9¹¹⁄₁₆
5 × 4	9¾	10½	1⁷⁄₁₆	11³⁄₁₆
6 × 2	6	8¼	2⁵⁄₁₆	8⁵⁄₁₆
6 × 3	7¹³⁄₁₆	9½	1¹⁵⁄₁₆	9¾
6 × 4	9¾	11	1⁷⁄₁₆	11³⁄₁₆
6 × 5	11¹¹⁄₁₆	13	¹³⁄₁₆	12½

Table 14-32 Single and Double Combination Y and $^1/_4$ Bends (mm)

Size	A	B	C	X
38	121	137	32	152
51	137	156	32	168
76	186	203	17	203
102	235	254	6	241
127	298	318	22	321
152	346	365	11	357
51 × 38	127	143	25	152
76 × 51	140	171	29	168
102 × 51	140	184	29	168
102 × 76	184	216	19	203
127 × 51	151	197	54	205
127 × 76	197	229	49	246
127 × 102	248	267	37	284
152 × 51	152	210	59	211
152 × 76	198	241	49	248
152 × 102	248	279	37	284
152 × 127	297	330	21	318

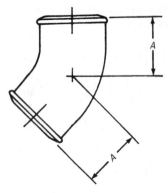

Fig. 14-37 Bends—1/6, 1/8, 1/16, 1/5.

Table 14-33 1/5 Bends

Size		A	
Inches	mm	Inches	mm
2	51	3 11/16	94
3	76	4 1/16	103
4	102	4 1/16	113

Table 14-34 1/6 Bends

Size		A	
Inches	mm	Inches	mm
2	51	3 1/4	83
3	76	3 1/2	89
4	102	3 13/16	97

Cast-Iron Pipe and Fittings 259

Table 14-35 $^1/_8$ Bends

Size		A	
Inches	mm	Inches	mm
$1^1/_2$	38	$2^5/_8$	67
2	51	$2^3/_4$	70
3	76	3	76
4	102	$3^1/_8$	79
5	127	$3^7/_8$	98
6	152	$4^1/_{16}$	103

Table 14-36 $^1/_{16}$ Bends

Size		A	
Inches	mm	Inches	mm
2	51	$2^1/_8$	54
3	76	$2^1/_4$	57
4	102	$2^5/_{16}$	59
5	127	$2^{15}/_{16}$	75
6	152	3	76

Sweeps

Fig. 14-38 shows sweeps, while Table 14-37 shows short-sweep specifications, Table 14-38 shows long-sweep specifications, and Table 14-39 shows reducing long-sweep specifications.

260 Chapter 14

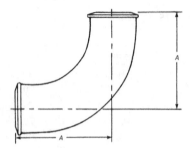

Fig. 14-38 Sweep.

Table 14-37 Short Sweeps

Size		A	
Inches	mm	Inches	mm
2	51	$6^1/_2$	165
3	76	7	178
4	102	$7^1/_2$	191
5	127	$8^1/_2$	216
6	152	9	229

Table 14-38 Long Sweeps

Size		A	
Inches	mm	Inches	mm
$1^1/_2$	38	$9^1/_4$	235
2	51	$9^1/_2$	241
3	76	10	254
4	102	$10^1/_2$	267
5	127	$11^1/_2$	292
6	152	12	305
8	203	$13^1/_2$	343

Cast-Iron Pipe and Fittings **261**

Table 14-39 Reducing Long Sweeps

Size		A	
Inches	mm	Inches	mm
3 × 2	76 × 51	10	254
4 × 3	102 × 76	10½	267

¹/₄ Bends

Fig. 14-39 shows a ¹/₄-bend fitting, and Table 14-40 shows ¹/₄-bend specifications.

Test Tees

Fig. 14-40 shows a test-tee fitting, and Table 14-41 shows test-tee specifications.

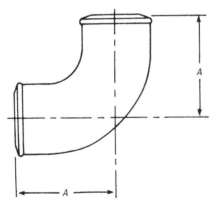

Fig. 14-39 ¹/₄ bend.

Table 14-40 ¹/₄ Bend

Size		A	
Inches	mm	Inches	mm
1¹/₂	38	4¹/₄	108
2	51	4¹/₂	114
3	76	5	127
4	102	5¹/₂	140
5	127	6¹/₂	165
6	152	7	178

Single and Double Sanitary T-Branches

Fig. 14-41 shows single and double sanitary T-branches, while Table 14-42 shows specifications for single and double sanitary T-branches in inches and Table 14-43 shows specifications for single and double sanitary T-branches in millimeters.

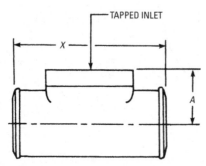

Fig. 14-40 Test tees.

Cast-Iron Pipe and Fittings 263

Table 14-41 Test Tees

Size		X		A	
Inches	mm	Inches	mm	Inches	mm
2	51	6³/₈	162	2	51
3	76	7³/₄	197	2¹¹/₁₆	68
4	102	8⁷/₈	225	3	76
5	127	11¹/₂	292	4¹/₂	114
6	152	23¹/₂	318	5	127

Single and Double Sanitary T-Branches, Tapped

Fig. 14-42 shows single and double sanitary T-branches, tapped, while Table 14-44 shows specifications for single and double sanitary T-branches, tapped, in inches and Table 14-45 shows specifications for single and double sanitary T-branches, tapped, in millimeters.

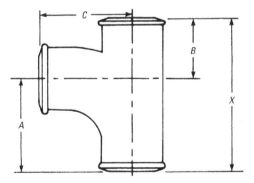

Fig. 14-41 Single and double sanitary T-branches.

Chapter 14

Table 14-42 Single and Double Sanitary T-Branches (Inches)

Size	A	B	C	X
1½	4¼	2¼	4¼	6½
2	4½	2⅜	4½	6⅞
3	5	3	5	8
4	5½	3⅝	5½	9⅛
3 × 1½	4¼	2¼	5	6½
3 × 2	4½	2⅜	5	6⅞
3 × 4	5½	3½	5	9
4 × 2	4½	2⅜	5½	6⅞
4 × 3	5	3	5½	8
5 × 2	5	3½	6½	8½

Table 14-43 Single and Double Sanitary T-Branches (mm)

Size	A	B	C	X
38	108	57	108	165
51	114	60	114	175
76	127	76	127	203
102	140	92	140	232
76 × 38	108	57	127	165
76 × 51	114	60	127	175
76 × 102	140	89	127	229
102 × 51	114	60	140	175
102 × 76	127	76	140	203
127 × 51	127	89	165	216

Cast-Iron Pipe and Fittings 265

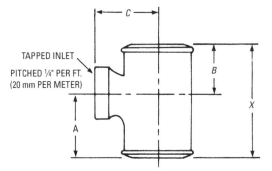

Fig. 14-42 Single and double sanitary T-branches, tapped.

Table 14-44 Single and Double Sanitary T-Branches, Tapped (Inches)

Size	A	B	C	X
$1\frac{1}{2} \times 1\frac{1}{4}$	$3\frac{1}{4}$	$2\frac{7}{16}$	$2\frac{9}{16}$	$5\frac{11}{16}$
$1\frac{1}{2} \times 1\frac{1}{2}$	$3\frac{1}{4}$	$2\frac{7}{16}$	$2\frac{9}{16}$	$5\frac{11}{16}$
$2 \times 1\frac{1}{4}$	$3\frac{1}{4}$	$2\frac{7}{16}$	$2\frac{13}{16}$	$5\frac{11}{16}$
$2 \times 1\frac{1}{2}$	$3\frac{1}{4}$	$2\frac{7}{16}$	$2\frac{13}{16}$	$5\frac{11}{16}$
2×2	$3\frac{3}{4}$	$2\frac{5}{8}$	$3\frac{1}{16}$	$6\frac{3}{8}$
$3 \times 1\frac{1}{4}$	$3\frac{1}{4}$	$2\frac{7}{16}$	$3\frac{5}{16}$	$5\frac{11}{16}$
$3 \times 1\frac{1}{2}$	$3\frac{1}{4}$	$2\frac{7}{16}$	$3\frac{5}{16}$	$5\frac{11}{16}$
3×2	$3\frac{3}{4}$	$2\frac{5}{8}$	$3\frac{9}{16}$	$6\frac{3}{8}$
$4 \times 1\frac{1}{4}$	$3\frac{1}{4}$	$2\frac{7}{16}$	$3\frac{13}{16}$	$5\frac{11}{16}$
$4 \times 1\frac{1}{2}$	$3\frac{1}{4}$	$2\frac{7}{16}$	$3\frac{13}{16}$	$5\frac{11}{16}$
4×2	$3\frac{3}{4}$	$2\frac{5}{8}$	$4\frac{1}{16}$	$6\frac{3}{8}$

Table 14-45 Single and Double Sanitary T-Branches, Tapped (mm)

Size	A	B	C	X
38 × 32	83	62	65	144
38 × 38	83	62	65	144
51 × 32	83	62	71	144
51 × 38	83	62	71	144
51 × 51	95	67	78	162
76 × 32	83	62	84	144
76 × 38	83	62	84	144
76 × 51	95	67	90	162
102 × 32	83	62	97	144
102 × 38	83	62	97	144
102 × 51	95	67	103	162

Y-Branches, Tapped
Fig. 14-43 shows Y-branches, tapped, while Table 14-46 shows specifications for Y-branches, tapped, in inches and Table 14-47 shows specifications in millimeters.

Upright Y-Branches
Fig. 14-44 shows upright Y-branches, while Table 14-48 shows specifications for upright Y-branches in inches and Table 14-49 shows specifications in millimeters.

$1/4$ Bend, Double
Fig. 14-45 shows a $1/4$ bend, double, and Table 14-50 shows specifications for a $1/4$ bend, double.

$1/4$ Bend, Short-Radius, Tapped
Fig. 14-46 shows a $1/4$ bend, short-radius, tapped, and Table 14-51 shows specifications for a $1/4$ bend, short-radius, tapped.

Cast-Iron Pipe and Fittings **267**

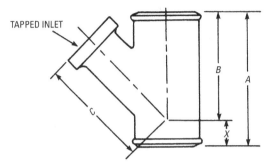

Fig. 14-43 Y-branches, tapped.

Table 14-46 Y-Branches, Tapped (Inches)

Size	A	B	C	X
2 × 1¼	6⅝	5	5¹⁄₁₆	1⅝
2 × 1½	6⅝	5	5¹⁄₁₆	1⅝
2 × 2	6⅝	4⅝	5¹⁄₁₆	2
3 × 1¼	6⅝	5⅛	5¹⁄₁₆	1½
3 × 1½	6⅝	5½	5¾	1⅛
3 × 2	6⅝	5⅛	5¹³⁄₁₆	1½
4 × 1¼	6⅝	5⅝	6⁷⁄₁₆	1
4 × 1½	6⅝	5⅝	6⁷⁄₁₆	1
4 × 2	6⅝	5⅝	6½	1

268 Chapter 14

Table 14-47 Y-Branches, Tapped (mm)

Size	A	B	C	X
51 × 32	168	127	129	41
51 × 38	168	127	129	41
51 × 51	168	117	129	51
76 × 32	168	130	129	38
76 × 38	168	140	146	29
76 × 51	168	130	148	38
102 × 32	168	143	164	25
102 × 38	168	143	164	25
102 × 51	168	143	165	25

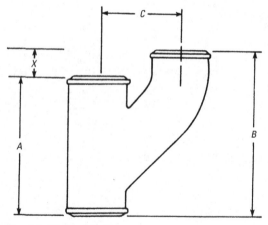

Fig. 14-44 Upright Y.

Cast-Iron Pipe and Fittings **269**

Table 14-48 Upright Y (Inches)

Size	A	B	C	X
2	7	$10^1/_4$	$5^1/_2$	$3^1/_4$
3	$8^5/_8$	$10^5/_8$	$5^1/_2$	$2^1/_4$
4	$9^3/_4$	$11^9/_{16}$	6	$1^{13}/_{16}$
3 × 2	7	$9^{11}/_{16}$	$5^1/_2$	$2^{11}/_{16}$
4 × 2	7	$9^1/_4$	$5^1/_2$	$2^1/_4$
4 × 3	$8^3/_8$	$10^1/_8$	$5^1/_2$	$1^3/_4$

Table 14-49 Upright Y (mm)

Size	A	B	C	X
51	178	260	140	83
76	213	270	140	57
102	248	294	152	46
76 × 51	178	246	140	68
102 × 51	178	235	140	57
102 × 76	213	257	140	44

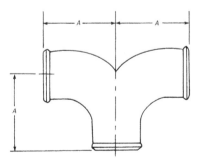

Fig. 14-45 $^1/_4$ bend, double.

Table 14-50 $^1\!/_4$ **Bend, Double**

Size		A	
Inches	mm	Inches	mm
2	51	$4^1\!/_2$	114
3	76	5	127
4	102	$5^1\!/_2$	140

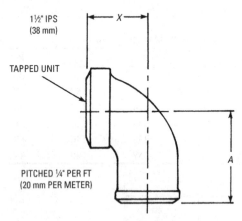

Fig. 14-46 Short-radius tapped $^1\!/_4$ bend.

Spread Between Vent and Revent When Using a Wye and $^1\!/_4$ Bend

Fig. 14-47 shows a spread between vent and revent when using a wye and $^1\!/_4$ bend, while Table 14-52 shows specifications for a spread between vent and revent when using a wye and $^1\!/_4$ bend in inches and Table 14-53 shows specifications in millimeters.

Cast-Iron Pipe and Fittings **271**

Table 14-51 Short-Radius Tapped $1/4$ Bend

Size		A		X	
Inches	Mm	Inches	mm	Inches	mm
$1 1/2 \times 1 1/4$	38×32	3	76	2	51
$1 1/2 \times 1 1/2$	38×38	3	76	2	51
$2 \times 1 1/4$	51×32	$3 1/4$	83	$2 1/4$	57
$2 \times 1 1/2$	51×38	$3 1/4$	83	$2 1/4$	57

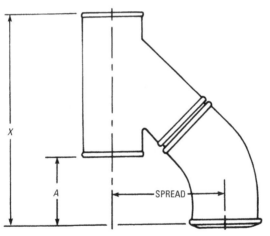

Fig. 14-47 Spread between vent and revent when using a wye and $1/4$ bend, to nearest $1/16$.

Table 14-52 Spread Between Vent and Revent with Wye (Inches)

Size	Spread	X	A
2 × 2	$5^5/_{16}$	$10^1/_{16}$	$3^7/_{16}$
3 × 3	$6^1/_4$	$11^1/_2$	$3^1/_2$
4 × 4	$7^1/_4$	$12^{12}/_{16}$	$3^5/_{16}$
3 × 2	$5^3/_4$	10	$3^3/_8$
4 × 2	$6^1/_4$	10	$3^3/_8$
4 × 3	$6^{13}/_{16}$	$11^1/_2$	$3^1/_2$
5 × 2	$7^5/_{16}$	11	$2^{15}/_{16}$
5 × 3	$7^7/_8$	$12^9/_{16}$	$2^7/_8$
5 × 4	$8^5/_{16}$	$13^7/_8$	$2^{11}/_{16}$
6 × 2	$7^7/_8$	$11^1/_8$	$2^{13}/_{16}$
6 × 3	$8^3/_8$	$12^5/_8$	$2^7/_8$
6 × 4	$8^{13}/_{16}$	$13^7/_8$	$2^{11}/_{16}$

Table 14-53 Spread Between Vent and Revent with Wye (mm)

Size	Spread	X	A
51 × 51	135	256	87
76 × 76	159	292	89
102 × 102	184	325	84
76 × 51	146	254	86
102 × 51	159	254	86
102 × 76	173	292	89
127 × 51	186	279	75
127 × 76	200	319	73
127 × 102	211	352	68
152 × 51	200	283	71
152 × 76	213	321	73
152 × 102	224	352	68

Cast-Iron Pipe and Fittings **273**

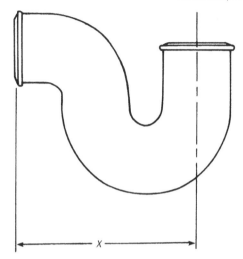

Fig. 14-48 ½ S- or P-traps.

½ S- or P-Traps
Fig. 14-48 shows ½ S- or P-traps, and Table 14-54 shows specifications for ½ S- or P-traps.

Tapped Extension Place
Fig. 14-49 shows a tapped extension place, and Table 14-55 shows specifications for a tapped extension place.

Increaser-Reducer
Fig. 14-50 shows an increaser-reducer, and Table 14-56 shows specifications for an increaser-reducer.

Minimum Offsets Using Bend Fittings
Fig. 14-51 shows minimum offsets for bends, while Table 14-57 shows specifications for ⅕ bends, Table 14-58 shows

274 Chapter 14

Table 14-54 $\frac{1}{2}$ **S- or P-Traps**

Size		X	
Inches	mm	Inches	mm
$1\frac{1}{2}$	38	$6\frac{3}{4}$	171
2	51	$7\frac{1}{2}$	191
3	76	9	229
4	102	$10\frac{1}{2}$	267
6	152	14	356

specifications for $\frac{1}{6}$ bends, Table 14-59 shows specifications for $\frac{1}{8}$ bends, and Table 14-60 shows specifications for $\frac{1}{16}$ bends.

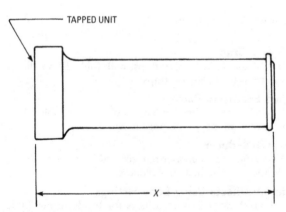

Fig. 14-49 Tapped extension place.

Cast-Iron Pipe and Fittings **275**

Table 14-55 Tapped Extension Place

Size		X		I.P.S. Tapping	
Inches	mm	Inches	mm	Inches	mm
2	51	12	305	2	51
3	76	12	305	3	76
4	102	12	305	$3^1/_2$	89

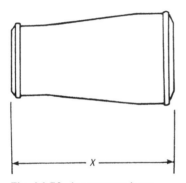

Fig. 14-50 Increaser-reducer.

Table 14-56 Increaser-Reducer

Size		X	
Inches	mm	Inches	mm
$1^1/_2 \times 2$	38 × 51	$3^5/_8$	92
2 × 3	51 × 76	8	203
2 × 4	51 × 102	8	203
3 × 4	76 × 102	8	203
4 × 6	102 × 152	4	102
5 × 6	127 × 152	$4^1/_2$	114

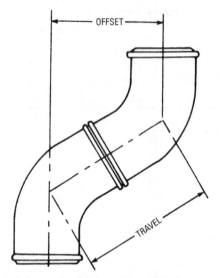

Fig. 14-51 Offset using bends.

Table 14-57 Minimum Offsets Using No-Hub $^1/_5$-Bend Fittings

Size		Travel		Minimum Offset	
Inches	mm	Inches	mm	Inches	mm
2	51	$7^{15}/_{32}$	190	$7^1/_{16}$	179
3	76	$8^7/_{32}$	209	$7^{13}/_{16}$	198
4	102	$8^{31}/_{32}$	228	$8^1/_2$	216

Note: Minimum inch offsets figured to nearest $^1/_{16}$ inch; minimum metric offsets to nearest millimeter.

Cast-Iron Pipe and Fittings **277**

Table 14-58 Minimum Offsets Using No-Hub $^1/_6$-Bend Fittings

Size		Travel		Minimum Offsets	
Inches	mm	Inches	mm	Inches	mm
2	51	$6^{19}/_{32}$	167	$5^{11}/_{16}$	144
3	76	$7^3/_{32}$	180	$6^1/_8$	156
4	102	$7^{23}/_{32}$	196	$6^{11}/_{16}$	170

Table 14-59 Minimum Offsets Using No-Hub $^1/_8$-Bend Fittings

Size		Travel		Minimum Offset	
Inches	mm	Inches	mm	Inches	mm
$1^1/_2$	38	$5^{11}/_{32}$	136	$3^{13}/_{16}$	97
2	51	$5^{19}/_{32}$	142	$3^{15}/_{16}$	100
3	76	$6^3/_{32}$	155	$4^5/_{16}$	110
4	102	$6^{11}/_{32}$	161	$4^1/_2$	114
5	127	$7^{27}/_{32}$	199	$5^9/_{15}$	141
6	152	$8^7/_{32}$	209	$5^{13}/_{16}$	148

Table 14-60 Minimum Offsets Using No-Hub $^1/_{16}$-Bend Fittings

Size		Travel		Minimum Offset	
Inches	mm	Inches	mm	Inches	mm
2	51	$4^{11}/_{32}$	110	$1^{11}/_{16}$	43
3	76	$4^{19}/_{32}$	117	$1^3/_4$	44
4	102	$4^{23}/_{32}$	120	$1^{13}/_{16}$	46
5	127	$5^{31}/_{32}$	152	$2^5/_{16}$	59
6	152	$6^3/_{32}$	155	$2^5/_{16}$	59

15. COPPER PIPE AND FITTINGS

This chapter provides an overview of copper pipe and fittings, as well as tips on using Sovent systems.

Types of Copper Tube

The two principal types of copper tube are plumbing tube and ARC tube. *Plumbing tube* includes types K, L, M, and DWV. *ARC tube* is used for air-conditioning and refrigeration. Table 15-1 lists the classes and types available.

Types K, L, M, and DWV come in standard sizes with the outside diameter always $\frac{1}{8}$ inch (0.32 cm) larger than the standard size. Each type represents a series of sizes with different wall thicknesses. Inside diameters depend on tube size and wall thickness. Drawn tube is hard tube, and annealed tube is soft. Hard tubing can be joined by soldering or brazing, by using capillary fittings, or by welding. Tube in the bending or soft tempers can be joined in the same ways, as well as with flare-type compression fittings. It is also possible to expand the end of one tube so that it can be joined to another by soldering or brazing without a capillary fitting.

Recommendations for Various Applications

Strength, formability, and other mechanical factors frequently determine the type of copper tube to be used in a particular application. Sometimes building or plumbing codes govern what types may be used. When a choice can be made, it is helpful to know which type of copper tube has served and will serve successfully and economically in the following applications:

- *Underground water services*—Use Type M for straight lengths joined with fittings, and Type L soft temper where coils are more convenient.

Table 15-1 Standard Copper Plumbing Tube Commercially Available Lengths

Tube	Drawn	Annealed
Type K Available in diameters from 1/4 inch to 12 inches (0.64 cm to 30.5 cm)	**Straight Lengths** Up to 8 inches (20.3 cm) S.P.S. 20 ft (6.1 m) 10 inches (25.4 cm) 18 ft (5.49 m) 12 inches (30.5 cm) 12 ft (3.66 m)	**Straight Lengths** Up to 8 inches (20.3 cm) 20 ft (6.1 m) 10 inches (25.4 cm) 18 ft (5.49 m) 12 inches (30.5 cm) 12 ft (3.66 m) **Coils** Up to 1 inch (2.54 cm) S.P.S. 60 ft (18.29 m) 100 ft (30.48 m) 1 1/4 inches and 1 1/2 inches (3.18 cm and 3.8 cm) 60 ft (18.29 m) 40 ft (12.19 m) 2 inches (5.1 cm) 45 ft (13.72 m)
Type L Available in diameters from 1/4 inch to 12 inches (0.64 cm to 30.5 cm)	**Straight Lengths** Up to 10 inches (25.4 cm) S.P.S. 20 ft (6.1 m) 12 inches (30.5 cm) 18 ft (5.49 m)	**Straight Lengths** Up to 10 inches (25.4 cm) 20 ft (6.1 m) 12 inches (30.5 cm) 18 ft (5.49 m) **Coils** Up to 1 inch (2.54 cm) 60 ft (18.29 m) 100 ft (30.48 m) 1 1/4 inches and 1 1/2 inches (3.18 cm and 3.8 cm) 60 ft (18.29 m) 2 inches (5.1 cm) 40 ft (12.19 m) 45 ft (13.72 m)

Type M Available in diameters from $3/8$ inch to 12 inches (0.95 cm to 30.5 cm)	**Straight Lengths** All diameters 20 ft (6.1 m)	**Straight Lengths** Up to 12 inches (30.5 cm) 20 ft (6.1 m) **Coils** Up to 1 inch (2.54 cm) 60 ft (18.29 m) 100 ft (30.48 m) $1^{1}/_{4}$ inches and $1^{1}/_{2}$ inches (3.18 cm and 3.8 cm) 60 ft (18.29 m) 2 inches (5.1 cm) 40 ft (12.19 m) 45 ft (13.72 m)
DWV Available in diameters from $1^{1}/_{4}$ inches to 8 inches (3.18 cm to 20.3 cm)	**Straight Lengths** All diameters 20 ft (6.1 m)	Not available
ARC Available in diameters from $1/6$ inch to $4^{1}/_{2}$ inches (0.32 cm to 10.48 cm)	**Straight Lengths** 20 ft (6.1 m)	**Coils** 50 ft (15.2 m)

- *Water distribution systems*—Use Type M for above and below ground.
- *Chilled water mains*—Use Type M for sizes up to 1 inch (2.54 cm) and Type DWV for sizes of $1\frac{1}{4}$ inches (3.18 cm) and larger.
- *Drainage and vent systems*—Use Type DWV for above- and below-ground, waste, soil, and vent lines, roof drainage, building drains, and building sewers.
- *Heating*—For radiant panel and hydronic heating and for snow-melting systems, use Type L soft temper where coils are formed in place or prefabricated, and use Type M where straight lengths joined with fittings are used. For hot-water heating and low-pressure steam, use Type M for sizes up to $1\frac{1}{4}$ inches (3.18 cm) and Type DWV for sizes of $1\frac{1}{4}$ inches and larger. For condensate return lines, Type L is successfully used.
- *Fuel oil and underground gas services*—Use copper tube in accord with local codes.
- *Oxygen systems*—Use Type L or K, suitably cleaned for oxygen service per the National Fire Protection Association (NFPA) of Boston, Massachusetts.

Installation Tips

This section provides tips on copper tube installation in the following areas:

- Small systems
- Pressure considerations
- Water demand

Small Systems

Distribution systems for single-family houses can be sized easily on the basis of experience and any applicable code

Table 15-2 Examples of Minimum Copper Tube Sizes for Short-Branch Connections to Fixtures

Fixture	Copper Tube Size, Inches (cm)
Drinking fountain	3/8 (0.95)
Lavatory	3/8 (0.95)
Water closet (tank type)	3/8 (0.95)
Bathtub	1/2 (1.27)
Dishwasher (home)	1/2 (1.27)
Kitchen sink (home)	1/2 (1.27)
Laundry tray	1/2 (1.27)
Service sink	1/2 (1.27)
Shower head	1/2 (1.27)
Sill cock, hose bibb, wall hydrant	1/2 (1.27)
Urinal (tank type)	1/2 (1.27)
Washing machine (home)	1/2 (1.27)
Kitchen sink (commercial)	3/4 (1.91)
Urinal (flush valve)	3/4 (1.91)
Water closet (flush valve)	1 (2.54)

requirements, as can other similar small installations. Detailed study of the six design considerations previously mentioned is not necessary in such cases. The size of the short branches to individual fixtures can be determined by reference to Table 15-2. In general, the mains servicing these fixture branches can then be sized as follows:

- Up to three 3/8-inch (0.95-cm) branches can be served by a 1/2-inch (1.27-cm) main.
- Up to three 1/2-inch (1.27-cm) branches or up to five 3/8-inch (0.95-cm) branches can be served by a 3/4-inch (1.91-cm) main.

- Up to three ¾-inch (1.91-cm) branches or correspondingly more ½-inch (1.27-cm) or ⅜-inch (0.95-cm) branches can be served by a 1-inch (2.54-cm) main.

Generous sizing within these limits will give good design and the best service. Working to minimum sizing within these guidelines will give an adequate system most of the time, depending on available main pressure and the probability of simultaneous use of fixtures. The water distribution system in many single-family homes with 2½ baths, for example, has been completely plumbed with copper tube of ¾-inch (1.91-cm) size and smaller. The sizing of more complex distribution systems requires detailed analysis of each of the size design considerations listed previously.

Pressure Considerations

The water service pressure at the point where the building distribution system (or segment or zone thereof) begins depends on the local main pressure, the local code, the pressure desired by the system designer, or a combination of these. In any case, it should not be higher than about 80 psi (552 kPa).

Water Demand

Each fixture in the system represents a certain demand for water. Some examples of approximate water demand in gallons per minute (gpm) or liters per minute (lpm) of flow are shown in Table 15-3.

Adding up numbers like these to cover all the fixtures in an entire building distribution system would give the total demand for water usage, in gallons per minute, if all the fixtures were operating and flowing at the same time (which, of course, does not happen). A reasonable estimate of demand is one based on the extent to which various fixtures in the building might actually be used simultaneously.

Table 15-3 Approximate Water Demand

Fixture	lpm	gpm
Drinking fountain	2.8	0.75
Lavatory faucet	7.6	2
Lavatory faucet, self-closing	9.5	2.5
Sink faucet, WC tank ball cock	11.4	3
Bathtub faucet, shower head, laundry tub faucet	15.1	4
Sill cock, hose bibb, wall hydrant	18.9	5
Flush valve (depending on design)	57–132	15–35

Pressure Losses Caused by Friction

The pressure available to move the water through the distribution system (or a part thereof) is the main pressure minus the following:

- The pressure loss in the meter
- The pressure needed to lift water to the highest fixture (static pressure loss)
- The pressure needed at the fixtures themselves

This remaining available pressure must be adequate to overcome the pressure losses caused by friction during flow of the total demand (intermittent plus continuous fixtures) through the distribution system and its various parts. The final operation is to select the tube sizes in accordance with the pressure losses caused by friction. In actual practice, the design operation may involve repeating the steps in the design process to readjust pressure, velocity, and size to achieve the best balance of main pressure, tube size, velocity, and available pressure at the fixtures for the design flow required in the various parts of the system.

Fig. 15-1 shows a collection of various types of fittings and adapters. Fig. 15-2 shows photos and drawings of specific drainage fittings.

Copper Sovent

The single-stack plumbing system, as the Sovent is called, was invented in 1959 by Fritz Sommer of Bern, Switzerland. It is a single-stack plumbing system designed to improve and simplify soil, waste, and vent plumbing in multistory buildings.

The basic design rules illustrated here are based on experience gained in the design and construction of hundreds of Sovent systems serving thousands of living units, not to mention the extensive experimental work conducted on the 10-story plumbing test tower.

The first Sovent installation was made in 1961 in Bern, Switzerland. Eight years later, 15,000 apartment units were installed in 200 buildings, up to 30 stories in height. Through the Copper Development Association, Inc., extensive tests were carried out on the instrumented test tower. Following the successful completion of these early tests, the system was brought to America.

The first installation was the Habitat Apartments constructed for Expo '67 in Montreal; the next, the Uniment structure in Richmond, California. In 1970, two large apartment buildings and a 10-story office building using Sovent were begun.

Each individual Sovent system must be designed to meet the conditions under which it will operate, and engineering judgment is required in applying the basic design rules presented here to specific buildings. The Copper Development Association, Inc., will be pleased to review Sovent system designs to help ensure that the design principles are followed.

Copper Pipe and Fittings **287**

Fig. 15-1 Assorted fittings and adapters. *(Courtesy NIBCO, Inc.)*

288 Chapter 15

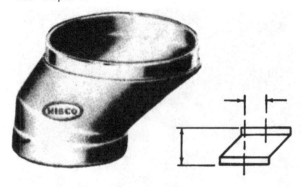

OFFSET CLOSET FITTING—FTGXC

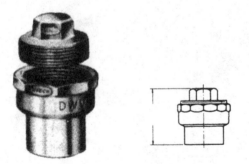

DWV—FTG. X CLEANOUT W/PLUG

Fig. 15-2 Copper drainage fittings.

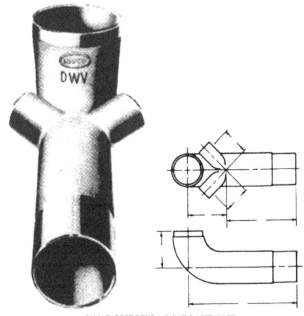

DWV CLOSET BEND—RIGHT & LEFT INLET

Fig. 15-2 (*continued*)

290 Chapter 15

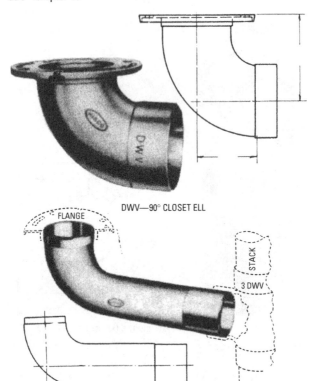

DWV—90° CLOSET ELL

DWV CLOSET BEND FTG. X FTG.

Fig. 15-2 (*continued*)

Copper Pipe and Fittings **291**

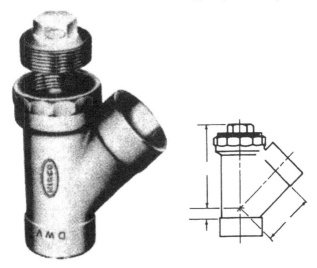

DWV—45° Y— WITH C.O. PLUG

DWV—45°—FTG. X COPPER ELL

Fig. 15-2 (*continued*)

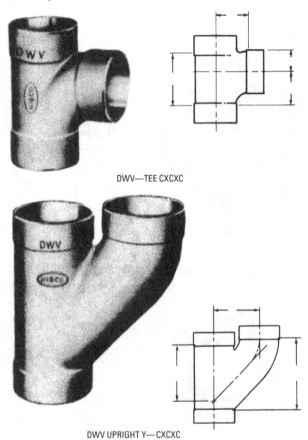

DWV—TEE CXCXC

DWV UPRIGHT Y—CXCXC

Fig. 15-2 (*continued*)

Copper Pipe and Fittings **293**

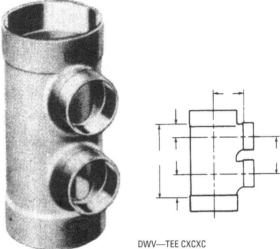

DWV—TEE CXCXC
DWV STACK FITTING—W/TWO SIDE INLETS

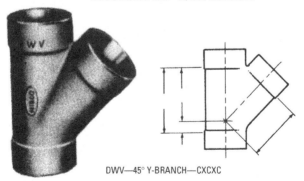

DWV—45° Y-BRANCH—CXCXC

Fig. 15-2 (*continued*)

294 Chapter 15

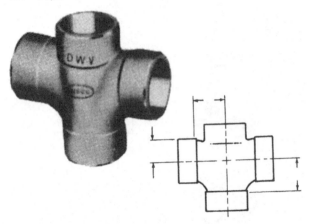

DWV VENT CROSS CXCXCXC

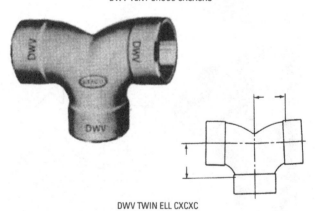

DWV TWIN ELL CXCXC

Fig. 15-2 (*continued*)

Copper Pipe and Fittings **295**

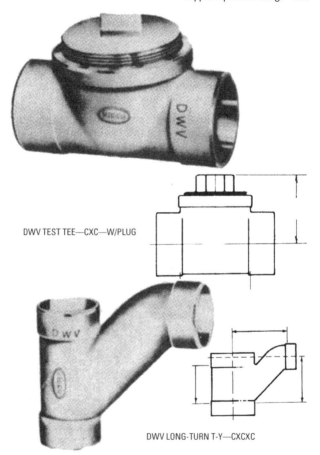

DWV TEST TEE—CXC—W/PLUG

DWV LONG-TURN T-Y—CXCXC

Fig. 15-2 (*continued*)

The copper Sovent plumbing system has four parts:

- A copper DWV stack
- A Sovent aerator fitting at each floor level
- Copper DWV horizontal branches
- A Sovent deaerator fitting at the base of the stack and at the upstream end of each horizontal offset

The two special fittings, the *aerator* and the *deaerator*, are the basis for the self-venting features of Sovent. It is claimed that a Sovent system will handle at least the same drainage fixture load as a conventional stack of the same diameter, but without the need for the separate vent stack and fixture revents needed in the traditional systems.

The aerator does three things:

- It limits the velocity of both liquid and air in the stack.
- It prevents the cross section of the stack from filling with a plug of water.
- It efficiently mixes the waste flowing in the branches with the air in the stack.

The Sovent aerator fitting is said to mix waste and air so effectively that no plug of water can form across the stack diameter and disturb fixture trap seals.

At a floor level where no aerator fitting is needed, a double in-line offset is used, as shown in Fig. 15-3.

Aerator Fittings

The Sovent aerator consists of an offset at the upper-stack inlet connection, a mixing chamber, one or more branch inlets, one or more waste inlets for connection of smaller waste branches, a baffle in the center of the chamber with an aperture between it and the top of the fitting, and the stack outlet at the bottom of the fitting. Waste branches connect to the side and face inlets on the aerator.

Copper Pipe and Fittings 297

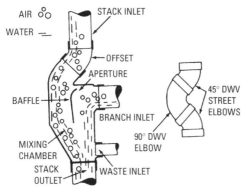

Fig. 15-3 Double in-line offset.

The aerator fitting provides a chamber where the flow of soil and waste from horizontal branches can unite smoothly with the air and liquid already flowing in the stack. The aerator fitting prevents pressure and vacuum fluctuations that could blow or siphon fixture trap seals.

No aerator fitting is needed at a floor level where no soil branch enters, and only a 2-inch (51-mm) waste branch enters a 4-inch (102-mm) stack. A double in-line offset is used in place of the aerator fitting. This offset reduces the fall rate in the stack between floor intervals in a manner similar to the aerator fitting (see Fig. 15-4).

The Sovent deaerator fitting relieves the pressure buildup at the bottom of the stack, venting that pressure into a relief line that connects into the top of the building drain. The deaerator pressure-relief line is tied to the building drain or at an offset to the lower stack (see Fig. 15-4 and Fig. 15-5).

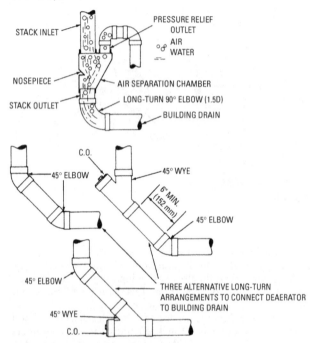

Fig. 15-4 Aerator fitting and long-turn details.

Copper Pipe and Fittings 299

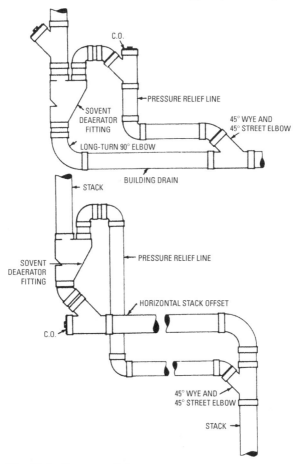

Fig. 15-5 Pressure-relief system.

Deaerator Fittings

The Sovent deaerator consists of an air-separation chamber having an internal nosepiece, a stack inlet, a pressure-relief outlet at the top, and a stack outlet at the bottom. The deaerator fitting at the bottom of the stack functions in combination with the aerator fitting above to make the single stack self-venting.

The deaerator is designed to overcome the tendency that would otherwise occur for the falling waste to build up excessive backpressure at the bottom of the stack when the flow is decelerated by the bend into the horizontal drain. The deaerator provides a method of separating air from system flow and equalizes pressure buildups.

Tests show that the configuration of the deaerator fitting causes part of the air falling with the liquid and solid in the stack to be ejected through the pressure-relief line to the top of the building drain, while the balance goes into the drain with the soil and waste.

At the deaerator outlet, the stack is connected into the horizontal drain through a long-turn fitting arrangement. Downstream at least 4 inches (122 cm) from this point, the pressure-relief line from the top of the deaerator fitting is connected into the top of the building drain.

A deaerator fitting, with its pressure-relief line connection, is installed in this way at the base of every Sovent stack and also at every offset (vertical-horizontal-vertical) in a stack. In the latter case, the pressure-relief line is tied into the stack immediately below the horizontal portion (see Fig. 15-6).

Stacks

The stack must be carried full size through the roof. Two stacks can be tied together at the top above the highest fixture, with one stack extending through the roof. If the distance between the two stacks is 20 feet (6.1 m) or less, the horizontal

Copper Pipe and Fittings 301

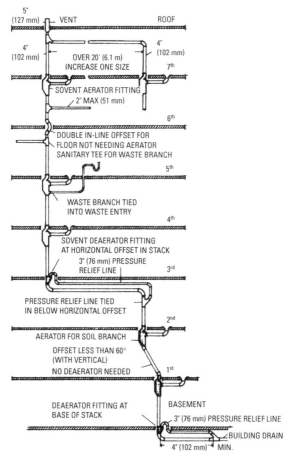

Fig. 15-6 Sovent drainage stack design features.

302 Chapter 15

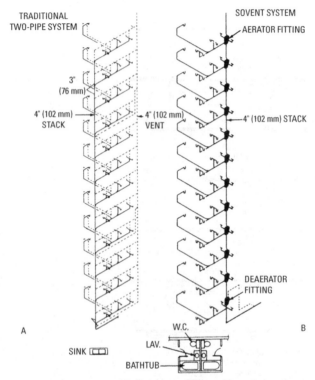

Fig. 15-7 Schematic drawing of a Sovent installation.

Copper Pipe and Fittings **303**

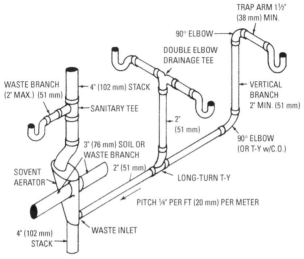

Fig. 15-8 Typical sewer and waste tie-ins.

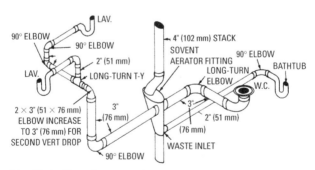

Fig. 15-9 Typical sewer and waste tie-ins.

304 Chapter 15

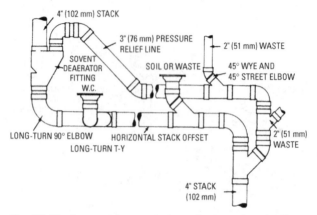

Fig. 15-10 Sewer and waste tie-ins to horizontal stack offset.

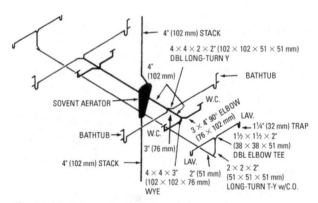

Fig. 15-11 Schematic of sewer and waste connections to a vertical stack.

Copper Pipe and Fittings **305**

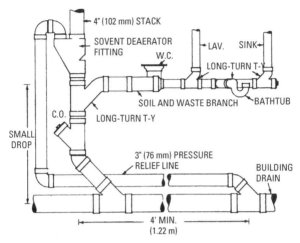

Fig. 15-12 Soil and waste connections just below deaerator at the bottom of the stack.

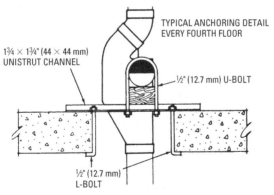

Fig. 15-13 Sovent stack anchoring.

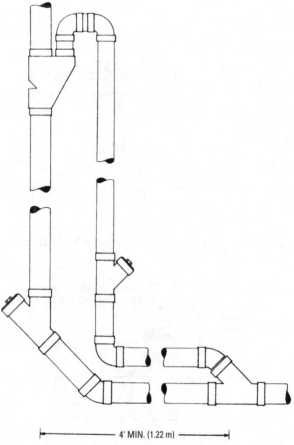

Fig. 15-14 Deaerator fitting above floor level of building drain.

Copper Pipe and Fittings **307**

Fig. 15-15 Installing a copper Sovent system.

308 Chapter 15

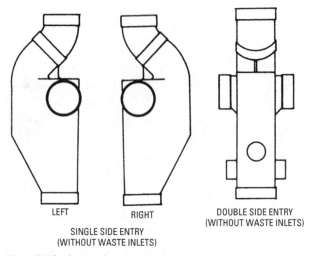

LEFT RIGHT
SINGLE SIDE ENTRY
(WITHOUT WASTE INLETS)

DOUBLE SIDE ENTRY
(WITHOUT WASTE INLETS)

Fig. 15-16 Sovent fittings with single- and double-side entries.

tie-line can be the same diameter as the stack that terminates below the roof level. If the distance is greater than 20 feet (6.1 m), the tie-line must be one size larger than the terminated stack.

The common stack extending through the roof must be one pipe size larger than the size of the larger stack below the tie-line.

The size of the stack is determined by the number of fixture units connected, as with traditional sanitary systems. Existing Sovent installations include 4-inch (102-mm) stacks serving up to 400 fixture units and 30 stories in height.

The Sovent's cost-saving potential can be seen by considering the illustration shown in Fig. 15-7 of a 12-story stack serving an apartment grouping. The material saving is shown graphically in the schematic riser diagrams for two-pipe and Sovent systems.

Sovent stacks are anchored in the same manner as others. Noise can be held to a minimum by making sure that no stack or branch is touching or bearing on a beam, brace, stud, or any other structural member. Sewer and waste lines can be tied, as shown in Fig. 15-8 and Fig. 15-9.

Soil and waste branches may be connected into a horizontal stack offset. Waste branches may be connected into the pressure-relief line, as shown in Fig. 15-10, Fig. 15-11, and Fig. 15-12.

Soil and waste branches may be connected immediately below a deaerator fitting at the bottom of the stack. The deaerator fitting may be located above the floor level of the building drain (see Figs. 15-13 through 15-16).

16. WATER HEATERS

This chapter provides useful tips on installation of traditional water heaters, as well as an examination of newer solar-powered water heaters.

Traditional Water Heaters

The DVE and the DRE models of the A.O. Smith Corp. are shown in Fig. 16-1. They have phase convertibility (that is, they can be changed over from single-phase to operate on a three-phase electrical power supply, and vice versa) in the field, without rewiring. Fig. 16-2 shows the gas-fired BTP model.

Location

A water heater is best located near a floor drain. Flushing and draining the tank is easier when there is a drain nearby. The discharge opening of the relief valve should always be piped to an open drain. In the interest of energy conservation, it is a good idea to keep hot-water piping and branch circuit wiring as short as possible and to insulate hot- and cold-water piping where heat loss and condensation may be a problem. The heater should be located in an area where leakage of the tank or connections will not result in damage to the area adjacent to the heater or to lower floors of the structure.

When such locations cannot be avoided, a suitable drain pan should be installed under the heater. The pan should be at least 2 inches greater than the length and width of the heater and should be piped to an adequate drain. Suggested clearances from adjacent surfaces are 18 inches (46 cm) in front for access to the controls and elements. The heater may be installed on or against combustible surfaces. The temperature of the space in which the heater is installed should not go below 32°F (0°C).

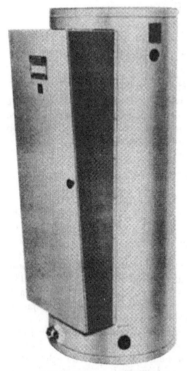

Fig. 16-1 The DVE and DRE Conservationists are energy-saving electric water heaters. The manufacturer reports that both can reduce energy costs by up to 43 percent and meet ASHRAE 90-75 Standard for energy efficiencies.

(Courtesy A.O. Smith Corporation, Consumer Products Division)

Water Heaters 313

Fig. 16-2 The BTP models of water heater are gas-fired.
(Courtesy A.O. Smith Corporation Consumer Products Division)

Water heater corrosion and component failure can be caused by the heating and breakdown of airborne chemical vapors. Spray can propellants, cleaning solvents, refrigerator and air-conditioning refrigerants, swimming pool chemicals, calcium and sodium chloride, waxes, and process chemicals are typical compounds that are potentially corrosive. These materials are corrosive at very low concentration levels with little or no odor to reveal their presence. Products of this sort should not be stored near the heater. Also, air brought into contact with the water heater should not contain any of these chemicals.

Installation

The heater water inlet is usually located on the side of the heater near the bottom. The heater outlet is located at the top of the heater. Fig. 16-3 shows a piping diagram for a two-temperature water heater with a mixing valve. Fig. 16-4 shows a one-temperature water heater.

An unplugged $\frac{3}{4}$-inch (1.91-cm) relief valve opening is provided for installing a listed temperature- and pressure-relief valve.

Install temperature- and pressure-protective equipment as required by local codes. The pressure setting of the relief valve should not exceed the pressure capacity of any component in the system. The temperature setting of the relief valve should not exceed 210°F (98.9°C).

Gas-fired water heaters may be installed on a combustible floor. However, clearance to adjacent surfaces must be provided as shown in Fig. 16-5. Units that are to be installed on combustible flooring must be supported by a full layer of hollow concrete blocks from 8 inches to 12 inches (20.32 cm to 30.48 cm) thick, and extending a minimum of 12 inches (30.48 cm) beyond the heater in all directions. The concrete blocks must provide an unbroken concrete surface under the heater, with the hollows running continuously and

Water Heaters 315

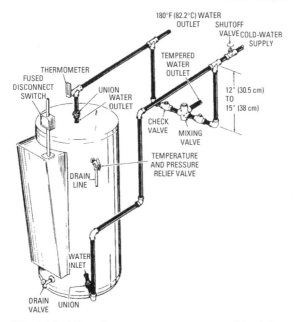

Fig. 16-3 Water heater—two temperature with mixing valve.
(Courtesy A.O. Smith Corporation, Consumer Products Division)

horizontally. If electrical conduits run around the floor of the proposed heater location, insulate the floor as recommended previously.

Some models are shipped with the burner installed on the heater. If the burner is shipped separately from the heater, it should be installed after the handling and leveling of the unit has been accomplished.

316 Chapter 16

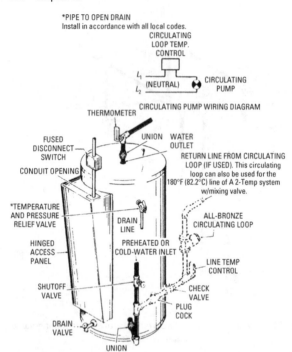

Fig. 16-4 Water heater—one temperature.
(Courtesy A.O. Smith Corporation, Consumer Products Division)

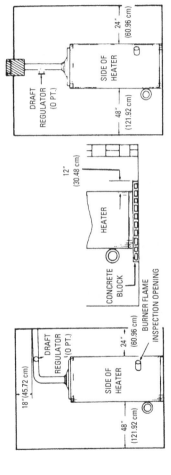

Fig. 16-5 Gas-fired heater installation. *(Courtesy A.O. Smith Corporation, Consumer Products Division)*

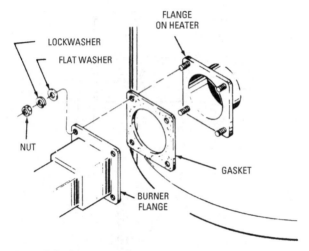

Fig. 16-6 Burner installation.
(Courtesy A.O. Smith Corporation, Consumer Products Division)

Assemble the burner and gaskets into the tank as shown in Fig. 16-6. The gas control string assembly may be shipped in the vertical position. When installing the unit, the gas control string must be in the horizontal position.

The water heater area should have sufficient air for satisfactory combustion of gas, proper venting, and maintenance of safe ambient temperature. The power burner maker's instructions should be followed with regard to the size of combustion and ventilation air openings. The opening size is based on the vent connector being equipped with a draft regulator. When the heater is installed in an area in which exhaust or ventilating fans may create unsatisfactory combustion or

Table 16-1 Vent Connector Diameter

BTP Model Number	Flue Outlet Inches (Centimeters)
200–300	6 (15.24)
200–600 & 400–600	8 (20.32)
200–800, 200–1000, 400–800, & 400–1000	10 (25.4)
200–1250, 200–1500, 400–1250, & 400–1500	12 (30.48)
400–1750 & 400–2000	14 (35.56)
400–2250 & 400–2500	16 (40.64)

venting, approved provisions must be made to overcome the problem.

The chimney vent connector diameter should be the same size as the heater flue outlet (see Table 16-1). A minimum rise of $1/4$ inch (0.64 cm) per foot (30.48 cm) of horizontal connector length must be maintained between the heater and chimney opening, as shown in Fig. 16-7. The connector length should be kept as short as possible.

Connectors must not be connected to a chimney vent or venting system served by a power exhauster unless the connection is made on the negative-pressure side of the exhauster. A draft regulator may be installed in the same room as the heater (see Fig. 16-7). Locate the regulator as close as possible to the heater and at least 18 inches (45.72 cm) from a combustible ceiling or wall. A manually operated damper should not be placed in the chimney connector.

Figs. 16-8 through 16-11 suggest typical methods of making water line connections to the heater. When a circulating pump is used, it is important to have a plug cock in the piping after the pump to regulate the water flow, preventing turbulence in the tank.

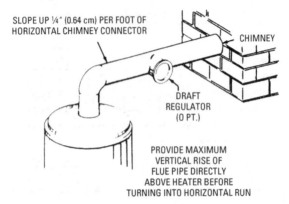

Fig. 16-7 Vent connection to a chimney.
(Courtesy A.O. Smith Corporation, Consumer Products Division)

The power burner maker's instructions should be followed with regard to the size and arrangement of the gas piping. The gas piping schematic shows field- and factory-installed piping and component locations at the power burner.

All electrical work must be installed in accordance with the *National Electrical Code* and local requirements. An electrical ground is required to reduce the risk of electrical shock. Do not energize the branch circuit before the heater tank is filled with water. A separate gas burner wiring diagram is supplied with the heater literature.

Following are steps to light the gas burner:

1. Turn the gas cock handle on the control to the "Off" position and dial the assembly to the lowest temperature position.

Water Heaters 321

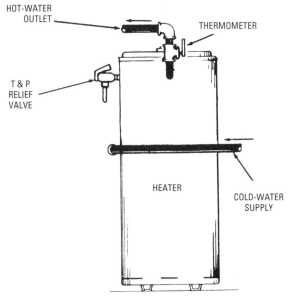

Fig. 16-8 Water line connection to a single-temperature model.
(Courtesy A.O. Smith Corporation, Consumer Products Division)

2. Wait approximately 5 minutes to allow gas that may have accumulated in the burner compartment to escape.
3. Turn the gas cock handle on the control to the "Pilot" position.
4. Fully depress the set button and light the pilot burner.
5. Allow the pilot to burn approximately 1 minute before releasing the set button. If the pilot does not remain lighted, repeat the operation.

322 Chapter 16

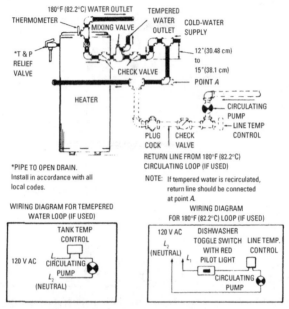

Fig. 16-9 Water line connection to a two-temperature, one-heater, high-temperature storage model, with or without recirculation. *(Courtesy A.O. Smith Corporation, Consumer Products Division)*

6. Turn the gas cock handle on the control to the "On" position and turn the dial assembly to the desired position. The main burner will then ignite. Adjust the pilot burner air shutter (if provided) to obtain a soft blue flame.

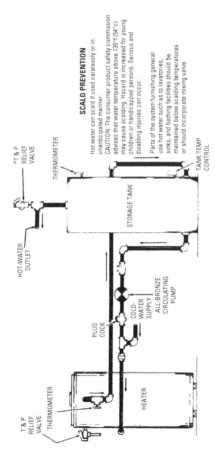

Fig. 16-10 Water line connections to a one-temperature, one-heater model with vertical storage tank forced circulation, with or without building recirculation.

(Courtesy A.O. Smith Corporation, Consumer Products Division)

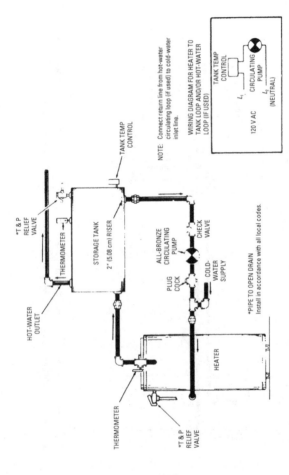

Fig. 16-11 Water line connections to a one-temperature, one-heater model with horizontal storage tank forced circulation, with or without building recirculation.
(Courtesy A.O. Smith Corporation, Consumer Products Division)

Water Heaters 325

Solar System Water Heaters

America's dwindling supply of energy is becoming more critical with each passing year. With the costs of electricity and gas rising, it may be wise to consider solar energy, or at least become familiar with a system that is probably here to stay.

The Conservationist solar system water heater shown in Fig. 16-12 can be tailored to most areas and prevailing conditions. A.O. Smith solar systems can be designed for use with existing hot-water heaters.

The Conservationist Solar System Water Heater

Here's how the A.O. Smith Conservationist solar system works (see Figs. 16-13 through 16-15).

The hot sun rays are absorbed by roof-mounted collector panels to heat special antifreeze fluid that is circulating through integral copper channels.

A closed-loop system is used for transfer of the heated solution and return. Propylene glycol eliminates any worries of freezing.

A heater-mounted differential controller has a modulating output to collect the maximum amount of available heat from collector panels, even on cloudy days. The pump is adjustable for flow with a restrictor that makes this solar system flexible for various installations.

A diaphragm expansion tank is provided on top of the heater to handle the expansion of heat-transfer fluid in a closed-loop circulating line.

Two high-density magnesium anodes protect the tank against corrosion.

A 3-inch double-efficiency blanket of high-density insulation surrounds the tank to keep in more heat.

The tank is isolated from the jacket to prevent conduction heat loss.

326 Chapter 16

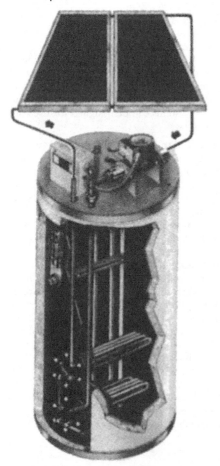

Fig. 16-12 The Conservationist solar system water heater.
(Courtesy A.O. Smith Corporation, Consumer Products Division)

Water Heaters 327

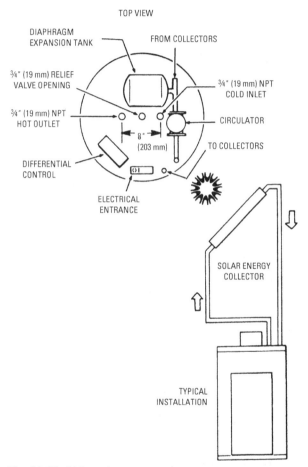

Fig. 16-13 Using solar energy to heat water.

328 Chapter 16

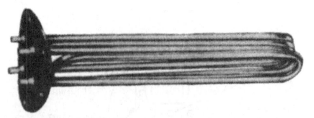

Fig. 16-14 Two elements in one unit—either or both may be operated to control how fast the heater can produce hot water.

Fig. 16-15 Single element life belt–type heating element.

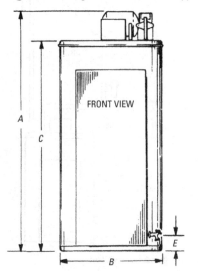

Fig. 16-16 Solar system water heater.

Table 16-2 Solar System Water Heater Specifications

	Model No.					
	Sun-82		Sun-100		Sun-120	
Size	Inches	Millimeters	Inches	Millimeters	Inches	Millimeters
A	56	1422	65⅞	1673	69	1753
B	28	711	28	711	30	762
C	48	1219	57⅞	1470	61	1549
E	4½	108	4½	108	5¾	146
Capacity	U.S. Gallons	Liters	U.S. Gallons	Liters	U.S. Gallons	Liters
	82	310.3	100	378.5	120	454.25
Approximate Shipping Weight	Pounds	Kilograms	Pounds	Kilograms	Pounds	Kilograms
	235	106.5	250	113.4	340	154.2

Exclusive Corona Heat Exchanger

The heat exchanger is immersed in the tank to ensure direct transfer of the heat. Ordinary exchangers are less efficient with a wraparound-tank method. The Corona has a double wall of copper for safety and is electrically isolated from the tank and external piping for positive protection against corrosion.

The Corona heat exchanger is used in Conservationist solar water heater models Sun-82, Sun-100, and Sun-120 gallon.

The Phoenix screw-in immersion element has two-way protection. A sheathing of iron-base *superalloy* provides excellent protection against burnout, oxidation, and scaling. The ceramic terminal block will not melt like ordinary plastic blocks. The Phoenix elements provide backup heating, as needed.

Specifications

Fig. 16-16 shows the front view of the solar system water heater, with letters designating measurement specifications corresponding to data in Table 16-2. The table also provides capacity and weight specifications.

17. WATER COOLERS AND FOUNTAINS

Water coolers are used in schools, industry, and hospitals. They require both plumbing and electrical attention. This chapter discusses installations of water coolers and fountains.

Features

Following are common features found in water coolers and drinking fountains (the numbers refer to those shown in Fig. 17-1):

1. The Dial-A-Drink Bubbler ensures a smooth, even flow of water under pressures from 20 psi to 125 psi (138 kPa to 862 kPa).

2. The stainless steel top of polished 18-8 stainless resists rust, corrosion, and stains. An antisplash ridge and integral drain direct and dispose of water.

3. The red brass cooling tank offers maximum cooling efficiency and reduces starts of the compressor. The 85-15 red brass storage tank (vented) has an internal heat-transfer surface and external refrigeration coil bonded to the tank by immersion in pure molten tin.

4. The copper cooling coils around the storage tank ensure maximum cooling efficiency. Double-wall separation of the refrigerant and drinking water conforms to all sanitary codes.

5. An insulating jacket of expanded polystyrene foam maintains cold-water temperature on all models.

6. An adjustable thermostat is tamper-proof. A remote-sensing bulb is located in the cooling tank and provides accurate control of cold-water temperature.

332 Chapter 17

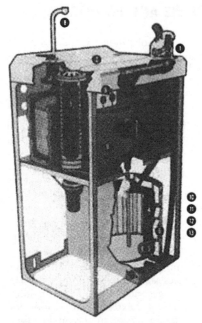

Fig. 17-1 Wall-hung water cooler.

7. The cost-cutting precooler (on larger capacity models) nearly doubles the capacity without extra operating cost by cooling incoming water with cold waste-water.

8. Hot water availability on Hot 'N Cold models: The hot tank heats and serves up to 45 cups of piping-hot water

per hour. The hot-water system is atmospherically vented and fiberglass-insulated.

9. The refrigeration system is maintenance free. The compressor and motor are hermetically sealed, lubricated for life, and leak-proof, if properly installed.
10. The durable cabinet finish includes vinyl laminatedonto steel on the front and side panels that provides a scuff-resistant finish.
11. ARI-certified performance means the cooling capacity of a water cooler is rated and certified in accordance with Air Conditioning and Refrigeration Institute (ARI) Standard 1010-73 (ANSI Standard A112-11.1-1973): gallons per hour of 50°F (10°C) drinking water with inlet temperature of 80°F (27°C) and room temperature of 90°F (32°C).
12. Most semi-recessed and simulated semi-recessed models have removable front and side panels, which provide extra work space for plumbing and electrical installations, servicing, and routine maintenance.
13. Each water cooler is performance tested.

Water Cooler Rough-In

Fig. 17-2 shows rough-in dimensions for a water cooler. Rough-in above water cooler is as follows:

- *Waste*—$22\frac{1}{2}$ inches (572 mm), $5\frac{1}{2}$ inches (140 mm) left of centerline.
- *Water*—$17\frac{1}{2}$ inches (445 mm), $6\frac{1}{2}$ inches (165 mm) left of centerline.
- *Water supply pipe*—$\frac{1}{2}$ inch (13 mm) N.P.S., reduced to an outside diameter of $\frac{3}{8}$ inch (10 mm); waste piping has an inside diameter of $1\frac{1}{4}$ inches (32 mm).

334 Chapter 17

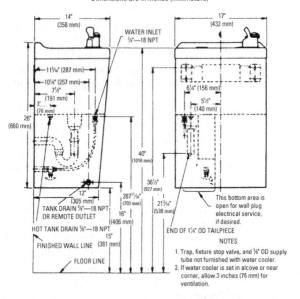

Fig. 17-2 Dimensions for mounting water cooler.

Number of People Served

MODEL	GHP of 50°	Offices	Light industry
ODP16M	15.7	188	100
ODP13M	13.0	156	91
ODP13M60	13.0	156	91
ODP7M	7.0	84	49
ODP7MH	7.0	84	49
ODP5M	5.0	60	35
ODPM	Nonrefrigerated Fountain		
ODP15MW	Water cooler		

Water Coolers and Fountains 335

Drinking Fountain Example

The Hastings Drinking Fountain shown in Fig. 17-3 is a familiar sight in schools, factories, and commercial locations. The drinking fountain is designed to have a complete water-saving, self-closing valve. The push-button operation makes for ease of cleaning and maintenance. The fountain is made of vitreous china and is shown with the elevated bubbler base, brass strainer, and wall hanger-mounted. The vandal-proof bubbler has the self-regulating valve and stop assembly.

Fig. 17-3 Vitreous china Hastings drinking fountain.
(Courtesy Eljer Plumbingware Co.)

336 Chapter 17

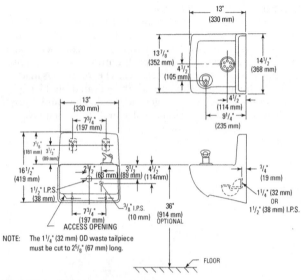

Fig. 17-4 Drinking fountain dimensions.
(Courtesy Eljer Plumbingware Co.)

This drinking fountain is furnished with a $\frac{3}{8}$-inch IPS supply connection and a $1\frac{1}{2}$-inch rough-cast brass P-trap. All roughing-in dimensions are provided in Fig. 17-4. All fixture dimensions are nominal and may vary within the range of tolerances established by the ASME/ANSI Standard A112.19.2M. The installer has the responsibility to comply with local codes and standards in the installation of the drinking fountain.

18. AUTOMATIC BATHROOM SYSTEMS

This chapter describes automatic flush valves used in water closets and urinals and provides installation instructions and troubleshooting information. The chapter concludes with a discussion of automatic shower heads.

The discussion in this chapter uses fixtures manufactured by Sloan Valve Company as examples. However, the tips and suggestions provided here may be adapted to other makes and models of fixtures.

Automatic Flush Valves

All Sloan flush valves made since 1906 can be repaired with parts (in kit form) obtained from the local plumbing supply house. It is recommended that the entire inside of the valve be replaced to make it appear new in operational characteristics. However, it is possible to replace the whole unit with an electronically controlled mechanism and obtain the added convenience of sanitary protection and automatic operation.

The Optima Plus uses advanced infrared technology to detect a user's presence and initiate a flushing cycle once the user steps away. The unit is powered by four AA batteries that will provide up to 3 years of service (based on 4000 flushes per month). They may also be powered by plug-in step-down transformers. The RESS models (see Fig. 18-1) are used to convert existing Royal and Regal style flushometers to sensor operation. The 8100 series Optima Plus valves (see Fig. 18-2) are complete flushometer valves and are ideal for new installations. When installing the valve, it is important that the flush model matches the requirements of the plumbing fixture. See Table 18-1.

Water Closet Flushometer

The Sloan exposed, battery-powered, sensor-operated G2 flushometer is designed for floor-mounted or wall-hung top spud bowls. It comes in Model 8111 for low-consumption (1.6 gpf) water closets and Model 8110 for water-saver (3.5 gpf) models (see Fig. 18-3). The control circuit is solid-state and operates on batteries configured to produce 6 volts input to the circuitry. There is an 8-second arming delay and a 3-second flush delay. The units use infrared to operate effectively with a sensor range of 22 to 42 inches (559 mm to 1067 mm) and an adjustable range of ± 8 inches. The power supply is composed of four AA batteries, preferably of the alkaline type. This way the batteries will last about 3 years at the rate of 4000 flushes per month. Indicator lights are the range adjustment and/or low-battery red light (it flashes when the batteries are low).

Fig. 18-1 Retrofit RESS series—conversion kit models.
(Courtesy Sloan Valve Company)

The unit will operate with 15 to 100 psi water pressure. The sentinel flush is set to flush once every 24 hours after the last flush. This keeps the traps filled and prevents odor caused by unflushed body wastes sitting in the bowl for extended periods.

Fig. 18-4 shows the rough-in for Model 8110/8111. Per the ADA Guidelines (Section 604.9.4), when installing the G2 Optima Plus in a handicap stall, it is recommended that the grab bars be split or shifted to the wide side of the stall. If grab bars must be present over the valve, use the alternate ADA installation shown in Fig. 18-5. This installation has

Automatic Bathroom Systems 339

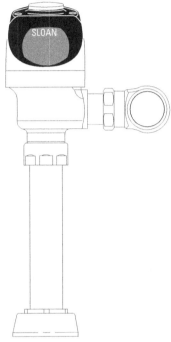

Fig. 18-2 8100 series complete flushometer model.
(Courtesy Sloan Valve Company)

a lower water supply rough-in to 10 inches (254 mm) and a mounted grab bar at the maximum allowed height of 36 inches (914 mm).

High rough-in water closet installation (see Fig. 18-6) is for Models 8113, 8115, and 8116. Models 8115 and 8116 are

Table 18-1 Optima Plus Water Closet and Urinal Models of Flushometers

Optima Plus Water Closet Models	Use
1.6 gpf/6.0 Lpf	For low-consumption bowls
2.4 gpf/9.0 Lpf	For 9-liter European water closets
3.5 gpf/13.2 Lpf	For older water closets
Optima Plus Urinal Models	**Use**
0.5 gpf/1.9 Lpf	For wash-down urinals
1.0 gpf/3.8 Lpf	For low-consumption urinals
1.5 gpf/5.7 Lpf	For older siphon-jet urinals
3.5 gpf/13.2 Lpf	For older blowout urinals

(Courtesy Sloan Valve Co.)

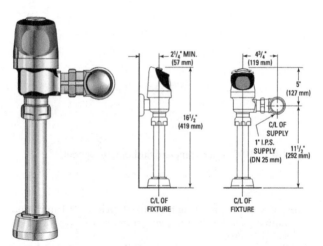

Fig. 18-3 Exposed, battery-operated, sensor-operated G2 model water closet flushometer for floor-mounted or wall-hung top spud bowls. *(Courtesy Sloan Valve Company)*

Automatic Bathroom Systems **341**

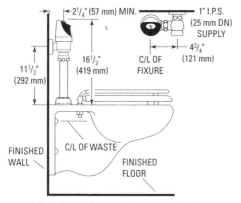

Fig. 18-4 Typical water closet installation.
(Courtesy Sloan Valve Company)

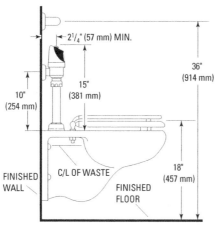

Fig. 18-5 Alternate ADA installation. *(Courtesy Sloan Valve Company)*

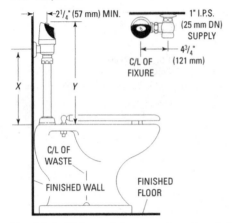

Fig. 18-6 High rough-in water closet installation.
(Courtesy Sloan Valve Company)

designed for installation where the water supply is roughed-in 24 inches to 27 inches (610 mm to 686 mm) above the top of the water closet.

For new installations, Sloan strongly recommends the use of their Model 8111, which has a shorter installation height. Use Model 8113 when toilet seats with covers are being used. The X and Y measurements for the valve installation are given in Table 18-2. For retrofit, make sure to specify RESS-C as the model being used.

Urinal Flushometer

The Sloan Model 8180 is the exposed, battery-powered, sensor-operated G2 model designed for urinal flushing (see Fig. 18-7). It also uses four AA dry cells for power that should last for 3 years at 4000 flushes per month. Also note that this model has an override switch mounted on top of the unit in

Automatic Bathroom Systems   343

Table 18-2 Rough-in Dimensions for Various Models of Flushometers

Model	X	Y
8113	16 inches (406 mm)	21 inches (533 mm)
8115	24 inches (610 mm)	29 inches (737 mm)
8116	27 inches (686 mm)	32 inches (813 mm)

(Courtesy Sloan Valve Co.)

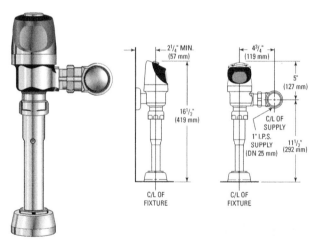

Fig. 18-7 Exposed, battery-powered, sensor-operated G2 model urinal flushometer. (Courtesy Sloan Valve Company)

344 Chapter 18

the form of a small push button. This unit can be obtained in polished brass, gold plate, brushed nickel, and satin chrome finishes to match any set of fixtures in any bathroom or restroom.

When used on a urinal, the 8180 can use low consumption (that is, 1 gpf or water-saver mode of 1.5 gpf). In the conventional configuration, it is used to flush the urinal with 3.5 gpf. The control circuitry and electrical specifications are the same as in earlier models of the G2 Optima Plus.

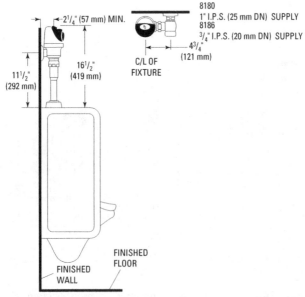

Fig. 18-8 Typical urinal installation using Models 8280 and 8186.
(Courtesy Sloan Valve Company)

Automatic Bathroom Systems **345**

Fig. 18-8 shows the rough-in for the G2 Optima Plus Model 8180. This is for new installations only. Install the closet or urinal fixture, drain line, and water supply line. It is important that all plumbing be installed in accordance with applicable codes and regulations. The water supply lines must be sized to provide an adequate volume of water for each fixture. Flush all water lines prior to making connections.

The G2 is designed to operate with 15 psi to 100 psi (104 kPa to 680 kPa) of water pressure. The minimum pressure required to the valve is determined by the type of fixture selected. Consult the fixture manufacturer for pressure requirements. Most low-consumption water closets (1.6-gallon/6-liter) require a minimum flowing pressure of 25 psi (172 kPa).

Typical Flushometer Installation

For complete valve installation, check the Internet for step-by-step installation instructions from the manufacturer (in this case, it would be Sloan Valve Company). Following are general installation steps:

1. **Install the control stop.** Install the Sloan Bak-Chek control stop to the water supply line with the outlet positioned as required (see Fig. 18-9). For sweat solder applications, refer to the following instructions and Fig. 18-10.

    - Measure the distance from the finished wall to the centerline of the fixture spud (see A in Fig. 18-10). Cut the water supply pipe $1\frac{1}{4}$ inches (32 mm) shorter than this measurement. Chamfer the OD and ID of the water supply pipe.

    - Slide the threaded adapter onto the water supply pipe until the end of the pipe rests against the shoulder of the adapter (see B in Fig. 18-10). Sweat-solder the adapter to the water supply pipe.

346 Chapter 18

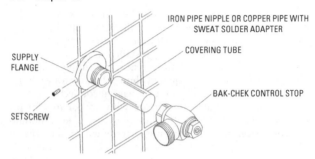

Fig. 18-9 Installing the control stop. *(Courtesy Sloan Valve Company)*

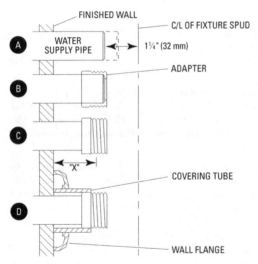

Fig. 18-10 Various installation possibilities.
(Courtesy Sloan Valve Company)

Automatic Bathroom Systems **347**

- Determine the length of the covering tube by measuring the distance from the finished wall to the first thread of the adapter (dimension X in C in Fig. 18-10). Cut the covering tube to this length.
- Slide the covering tube onto the water supply pipe. Slide the wall flange over the covering tube until it rests against the finished wall (see D in Fig. 18-10).
- Install the Sloan Bak-Chek control stop to the water supply line with the outlet positioned as shown in Fig. 18-9.

2. **Flush out the supply line.** Open the control stop (see Fig. 18-11). Turn on the water supply to flush the line of any debris or sediment. Close the control stop.

3. **Install the vacuum breaker flush connection.** Slide the spud coupling, nylon slip gasket, rubber gasket, and spud flange over the vacuum breaker tube (see Fig. 18-12). Insert the vacuum breaker tube into the fixture spud. Hand-tighten the spud coupling onto the fixture spud. If cutting the vacuum breaker tube to size, note that the critical line (C/L) on the vacuum breaker must typically be 6 inches (152 mm) above the fixture. Consult local codes for details.

4. **Install the flushometer.** The Sloan adjustable tailpiece (see Fig. 18-13) compensates for the off-center roughing-in on the job. Maximum

Fig. 18-11 Flushing out the supply line and opening the control stop with screwdriver.

(Courtesy Sloan Valve Company)

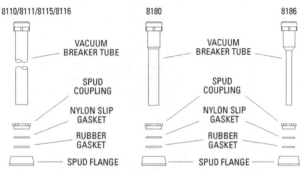

Fig. 18-12 Installing the vacuum breaker flush connection.
(Courtesy Sloan Valve Company)

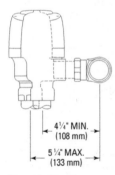

Fig. 18-13 Sloan adjustable tailpiece.
(Courtesy Sloan Valve Company)

adjustment is $\frac{1}{2}$ inch (13 mm) *in* or $\frac{1}{2}$ inch (13 mm) *out* from the standard $4\frac{3}{4}$-inch (121-mm) centerline of the flushometer to the centerline of the control stop (see Fig. 18-14). With the exception of the control stop inlet, *do not use pipe sealant* or *plumbing grease on any valve* or *Optima Plus component*.

Note

When retrofitting an existing valve, start here:

5. Remove components from existing flushometer. This step is for RESS retrofit installations only (see Fig. 18-15).

Automatic Bathroom Systems **349**

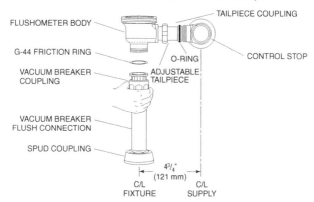

Fig. 18-14 Inserting adjustable tailpiece into the control stop.
(Courtesy Sloan Valve Company)

- Remove the control stop cap (see A in Fig. 18-15).
- Turn off the water supply at the control stop. Push the valve handle to relieve water pressure (see B in Fig. 18-15).
- Remove the outside and inside covers and old inside parts kit (see C in Fig. 18-15).
- Remove the old handle assembly and gasket (see D in Fig. 18-15).

Note
An extra H-533 tail O-ring is included in the event that leakage occurs when the valve is repositioned during the installation of the new Optima Plus. Use only if needed.

- Install the handle cap with gasket to the handle opening on the flushometer body (see E in Fig. 18-15). Tighten

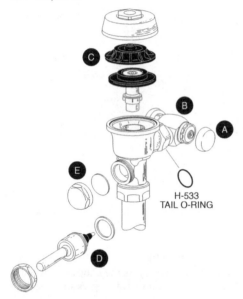

Fig. 18-15 Removing components from existing flushometer.
(Courtesy Sloan Valve Company)

the chrome handle cap securely. Keep in mind that this step is required only for retrofit installations.

6. **Install the flush volume regulator for G2, RESS-C, and RESS-U retrofit models.** The flush volume of the Optima Plus is controlled by the regulator in the flex tube diaphragm kit. Regulators are identified by color (see Table 18-3). The 0.5 gpf (1.9 Lpf) urinal kit can be converted to a 1.0 gpf (3.8 Lpf) by cutting and removing the smooth A-164 flow ring from the guide.

Automatic Bathroom Systems **351**

Table 18-3 Regulator Colors

Fixture and Flush	Regulator Color
0.5 gpf (1.9 Lpf) urinal	Green
1.0 gpf (3.8 Lpf) urinal	Green
1.5 gpf (5.7 Lpf) urinal	Black
1.6 gpf (6.0 Lpf) closet	Green
3.5 gpf (13.2 Lpf) closet	White
4.5 gpf (17.0 Lpf) closet	White
3.5 gpf (13.2 Lpf) urinal	White
2.4 gpf (9.0 Lpf) closet	Blue

When installing a new regulator on a flex tube diaphragm kit, be sure to push the regulator past the O-ring when installing (see Fig. 18-16). Note also that you never use more water than needed. Low-consumption water closets and urinals will not function properly on excess water.

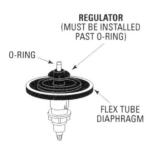

Fig. 18-16 Flex tube diaphragm, O-ring, and regulator.
(Courtesy Sloan Valve Company)

7. **Assemble the flex tube diaphragm in the Optima Plus assembly.** Remove the tab at the underside of the Optima Plus assembly (see Fig. 18-17). Insert the metal end of the flex tube into the hole at the underside of the Optima Plus assembly. The O-ring must be fully inserted into the hole. Push the diaphragm

352 Chapter 18

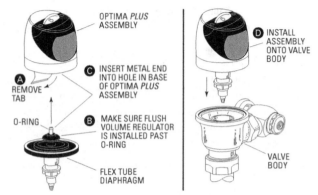

Fig. 18-17 Assembling the flex tube diaphragm to the Optima Plus assembly. *(Courtesy Sloan Valve Company)*

securely against the underside of the Optima Plus assembly. Place the entire assembly onto the valve body. Thread the locking ring into the valve body, and use the strap wrench provided to tightly secure the locking ring (see Fig. 18-18).

8. **Remove the tab to activate the sensor module.** Remove the tab located over the override button to activate the sensor module (see Fig. 18-19). For the first 10 minutes of operation, a visible red light flashes in the sensing window of the Optima Plus flushometer when a user is detected.

9. **Test the sensor for proper operation.** Stand in front of the Optima Plus sensor, wait 10 seconds, and then step away; the solenoid will click, indicating that the unit is operating (see Fig. 18-20).

Automatic Bathroom Systems **353**

Fig. 18-18 Tightening the locking ring. *(Courtesy Sloan Valve Company)*

Fig. 18-19 Removing the tab from the sensor module to activate the unit. *(Courtesy Sloan Valve Company)*

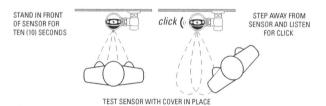

Fig. 18-20 Testing the sensor operation. *(Courtesy Sloan Valve Company)*

354 Chapter 18

The Optima Plus has a factory-set sensing range. For water closet models, it is 22 inches to 42 inches (559 mm to 1067 mm). For urinal models, the range adjustment is from 15 inches to 30 inches (381 mm to 762 mm). The factory setting should be satisfactory for most installations.

10. **Adjust the control stop to suit the fixture.** Adjust the control stop to meet the flow rate required for the proper cleansing of the fixture (see Fig. 18-21).

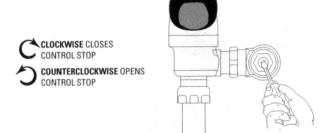

CLOCKWISE CLOSES CONTROL STOP

COUNTERCLOCKWISE OPENS CONTROL STOP

Fig. 18-21 Adjusting the control stop to suit the fixture with screwdriver. *(Courtesy Sloan Valve Company)*

11. **Install the stop cap.** For RESS retrofit applications, reuse the stop cap from the existing valve. In complete valve installation, a new stop cap is provided.

Operation

A continuous, invisible light beam is emitted from the sensor (see Fig. 18-22). As the user enters the beam's effective range, the beam is reflected into the scanner window to activate the output circuit. Once activated, the output circuit continues

Automatic Bathroom Systems **355**

Fig. 18-22 Checking operation of the sensor at the water closet toilet and the urinal. *(Courtesy Sloan Valve Company)*

in a hold mode for as long as the user remains within the effective range of the sensor. When the user steps away, the loss of reflected light initiates a one-time electrical signal that activates the flushing cycle to flush the fixture. The circuit automatically resets and is ready for the next user.

Fig. 18-23 shows the operation of the G2. Note the positioning of the light beam for effective utilization of the flushing process.

Operation of the 8180 unit is shown in the three steps described in Fig. 18-24. Note that the lens deflector is no longer needed for targeting children or wheelchair users.

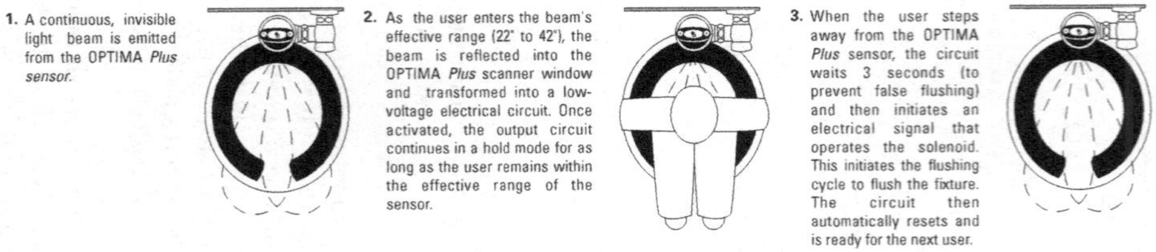

1. A continuous, invisible light beam is emitted from the OPTIMA *Plus* sensor.

2. As the user enters the beam's effective range (22" to 42"), the beam is reflected into the OPTIMA *Plus* scanner window and transformed into a low-voltage electrical circuit. Once activated, the output circuit continues in a hold mode for as long as the user remains within the effective range of the sensor.

3. When the user steps away from the OPTIMA *Plus* sensor, the circuit waits 3 seconds (to prevent false flushing) and then initiates an electrical signal that operates the solenoid. This initiates the flushing cycle to flush the fixture. The circuit then automatically resets and is ready for the next user.

Fig. 18-23 Operation of the water closet flushometer. (*Courtesy Sloan Valve Company*)

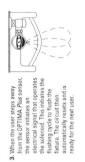

1. A continuous, invisible light beam is emitted from the OPTIMA *Plus* sensor.

2. As the user enters the beam's effective range (15" to 30"), the beam is reflected into the OPTIMA *Plus* scanner window and transformed into a low-voltage electrical circuit. Once activated, the output circuit continues in a hold mode for as long as the user remains within the effective range of the sensor.

3. When the user steps away from the OPTIMA *Plus* sensor, the sensor initiates an electrical signal that operates the solenoid. This initiates the flushing cycle to flush the fixture. The circuit then automatically resets and is ready for the next user.

Fig. 18-24 Operation of the urinal flushometer. *(Courtesy Sloan Valve Company)*

Range Adjustment

Adjustment of the range is performed only if necessary. The G2 Optima Plus has a factory-set sensing range that should be satisfactory for most installations. If the range is too short (that is, not properly sensing users) or too long (that is, sensing the opposite wall or stall door), the range can be adjusted.

Note
Water does not have to be turned off to adjust the range.

Loosen the two screws on top of the unit. Remove the override button. Remove the rubber plug from the top of the electronic sensor module to uncover the potentiometer (see Fig. 18-25).

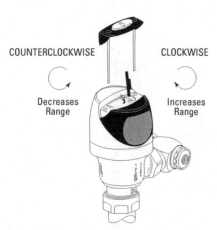

Fig. 18-25 Range adjustment for the G2 flushometer.
(Courtesy Sloan Valve Company)

Automatic Bathroom Systems **359**

For the first 10 minutes of operation, a visible red light flashes in the sensing window of the G2 when a user is detected. This visible red light feature can be reactivated after 10 minutes by opening and closing the battery compartment door.

Check the range by stepping toward the unit until the red light flashes, indicating the sensor's maximum detection limit. Adjust the range potentiometer screw located on top of the sensor module a few degrees clockwise to increase the range or a few degrees counterclockwise to decrease the range (see Fig. 18-25). Repeat this adjustment until the desired range is achieved. Always determine the sensing range with the metal cover and lens window on top of the unit.

Important
Adjust in small increments only! The range potentiometer adjustment screw rotates only $^3/_4$ of a turn. *Do not* over-rotate.

Battery Replacement
When required, replace the batteries with four alkaline type AA batteries exactly as shown in Fig. 18-26 and Fig. 18-27.

Note
Water does not have to be turned off to replace the batteries.

Install the battery compartment cover and secure with retaining screw. Make certain that the battery compartment cover is fully compressed

Fig. 18-26 Replacing batteries in the flushometer.
(Courtesy Sloan Valve Company)

360 Chapter 18

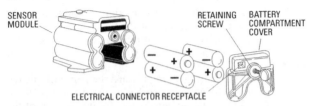

Fig. 18-27 Making sure the polarity of the cells is correct.
(Courtesy Sloan Valve Company)

against the gasket to provide a seal (*do not over-tighten*). Plug the electrical connector into the battery compartment cover. Reinstall the sensor module onto the plate. Install the complete cover assembly onto the plate. Tighten the two screws on top of the unit.

Care and Cleaning of Chrome and Special Finishes
Do not use abrasive or chemical cleaners to clean flushometers because they may dull the luster and attack the chrome or special decorative finishes. Use only soap and water, and then wipe dry with a clean cloth or towel.

While cleaning the bathroom tile, the flushometer should be protected from any splattering of cleaner. Acids and cleaning fluids can discolor or remove the chrome plating.

Troubleshooting
As with any appliance or unit with an electronic element, there is the possibility of incorrect operation from time to time. There are a few problems associated with the smooth operation of any automatic device. Keeping a close eye on the way it should be operating and the way it malfunctions will give you a clue as to what is wrong. The manufacturer provides a complete set of troubleshooting instructions and tables on the Internet.

Table 18-4 provides a guide to common problems, causes, and solutions.

Note
> The EBV-46-A beam deflector is no longer required or available for the G2 Optima Plus.

Act-O-Matic Shower Head

The Sloan Act-O-Matic self-cleaning, wall-mounted, shower head with adjustable spray direction is designed for institutional use. The chrome-plated shower head (see Fig. 18-28) has the following features. It is spring-loaded and self-cleaning with a spray disc that prevents particle clogging and a cone-within-a-cone spray pattern for total body coverage. It has an adjustable spray angle and pressure-compensating (2.5 gpm/9.4 Lpm) flow control, and it is made of all-brass construction. The mounting plate is vandal-resistant with its mounting screws. The water inlet is a $\frac{1}{2}$-inch-IPS pipe nipple with a shower head that is in conformance with all requirements of the ANSI/ASME Standard A112.18.1M, CSA B-125, and the United States Federal Energy Policy Act.

Prior to installation, there are a few things of which the plumber should be aware. The shower enclosure should be installed with a drain line and water supply line ($\frac{1}{2}$ inch IPS, or 13 mm DN) that have male thread for connection to the shower head.

Important
> All plumbing is to be installed in accordance with applicable codes and regulations. Flush all water lines until the water is clear before installing the shower head. The water lines must be sized to provide an adequate flow of water for each shower head. A minimum water pressure of 15 psi (103 kPa) is required.

Table 18-4 Troubleshooting Guide

Problem	Cause	Solution
Sensor flashes continuously (only when user steps within range).	Unit is in startup mode.	No problem. This feature is active for the first 10 minutes of operation.
Valve does not flush; sensor not picking up user.	Range too short.	Increase the range.
Valve does not flush; sensor picking up opposite wall or surface, or only flushes when someone walks by; red light flashes continuously for first 10 minutes, even with no one in front of the sensor.	Range too long.	Shorten the range.
Valve does not flush even after adjustment.	Range adjustment potentiometer set at full max or full min setting.	Readjust potentiometer away from full max or min setting.
	Batteries completely used up.	Replace batteries.
	Problem with the electronic sensor module.	Replace the module.

Problem	Cause	Solution
Unit flashes four quick times when user steps within range.	Batteries low.	Replace batteries.
Valve does not shut off.	Bypass orifice in diaphragm is clogged with dirt or debris, or bypass is clogged by an invisible gelatinous film caused by overtreated water. **Note:** The size of the orifice in the bypass is of utmost importance for the proper metering of water by the valve. Do not enlarge or damage this orifice. Replace the flex tube diaphragm if cleaning does not correct the problem. Dirt or debris fouling stem of the flex tube diaphragm.	Remove the flex tube diaphragm and wash under running water.
	O-ring on stem of flex tube diaphragm is damaged or worn.	Remove the flex tube diaphragm and wash under running water. Replace the O-ring if necessary.

(*continued*)

Table 18-4 (continued)

Problem	Cause	Solution
	Problem with electronic sensor module.	Replace the sensor module.
Not enough water to fixture.	Wrong flush volume regulator installed in the flex tube diaphragm kit.	Install the correct regulator.
	Wrong Optima Plus model installed (for example, a 1-gpf urinal installed on 3.5-gal closet fixture).	Replace with the proper Optima Plus model or convert existing unit to the proper model.
	Enlarged bypass in diaphragm.	Replace the flex tube diaphragm.
	Control stop not adjusted properly.	Readjust the control stop.
	Inadequate volume or pressure at supply.	Increase the water pressure or supply (flow) to the valve. Consult the factory for assistance.

Too much water to fixture.	Wrong flush volume regulator installed in the flex tube diaphragm kit.	Install the correct regulator.
	Control stop not adjusted properly.	Readjust the control stop.
	Wrong Optima Plus model installed (for example, a 3-gpf model installed for 1.0- or 1.5-gal urinal fixture).	Replace with proper Optima Plus model, or convert existing unit to the proper model.
	Dirt in diaphragm bypass.	Clean under running water or replace the flex tube diaphragm.

366 Chapter 18

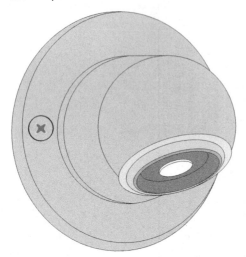

Fig. 18-28 The self cleaning, wall-mounted shower head with adjustable spray direction and integral 2.5-gpm flow control.
(Courtesy Sloan Valve Company)

Installation

Following are five steps for installing the Act-O-Matic shower head:

1. Locate the shower head mounting holes shown in Fig. 18-29. Drill four mounting holes $5/16$ inch $\times$ $1\frac{1}{2}$ inches (8 mm $\times$ 38 mm). Drive the anchors flush with the wall surface.
2. Install the mounting plate.
3. Install the pipe fitting as shown in Fig. 18-30.
4. Mount the shower head assembly (see Fig. 18-30).
5. Adjust the direction of the stream (see Fig. 18-31).

Automatic Bathroom Systems **367**

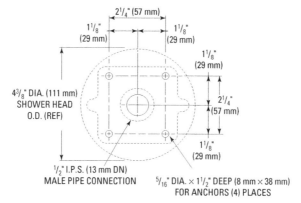

Fig. 18-29 Drill mounting holes for the shower head.
(Courtesy Sloan Valve Company)

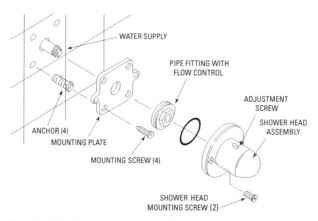

Fig. 18-30 How the shower head assembly is mounted.
(Courtesy Sloan Valve Company)

368 Chapter 18

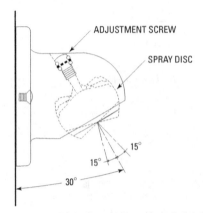

Fig. 18-31 Adjusting direction of the shower head stream.
(Courtesy Sloan Valve Company)

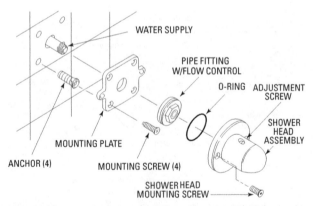

Fig. 18-32 Shower head installation (Model AC450).
(Courtesy Sloan Valve Company)

Automatic Bathroom Systems 369

PARTS LIST

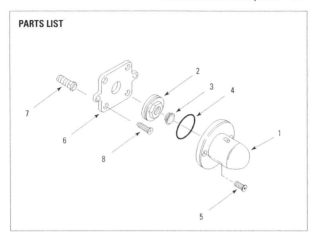

Item No.	Part No.	Description
1	AC-450	Shower Head Assembly
2 *	SH-511	Pipe Fitting
3 *	SH-1005-A	Flow Control, 2.5 gpm (9.4 Lpm)
4 *	SH-512	O-Ring
5	SH-514	Shower Head Mounting Screw (2)
6	SH-510	Mounting Plate
7	K-57	Anchor (4)
8	SH-37	Mounting Screw (4)

* Items 2, 3, and 4 are supplied assembled.

Fig. 18-33 Parts list for the shower head.
(Courtesy Sloan Valve Company)

370 Chapter 18

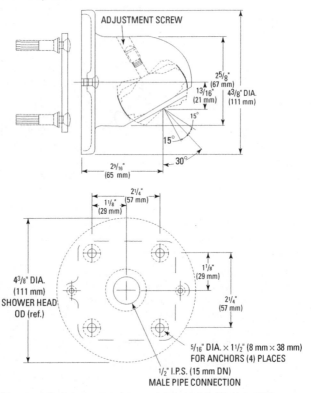

Fig. 18-34 Shower head measurements (Model AC-450).
(Courtesy Sloan Valve Company)

Repair Kit and Parts List

For a shower repair kit for the AC-450 Act-O-Matic shower head, use the designation SH-1009-A (Code # 4328471). It includes a spray disc assembly, a spray disc O-ring, an adjustable screw O-ring, a pipe fitting O-ring, and a 2.5-gpm (9.4-Lpm) flow control. An exploded view of the shower head and its alignment for installation is shown in Fig. 18-32. See the parts list in Fig. 18-33. Fig. 18-34 gives a view of the actual measurements associated with the installation.

Part III
General Reference Information

19. ABBREVIATIONS, DEFINITIONS, AND SYMBOLS

This chapter provides a reference for abbreviations, definitions, specifications, and symbols commonly used by plumbers.

Abbreviations
Following are some common abbreviations:

- AGA—American Gas Association
- ANMC—American National Metric Council
- ASA—American Standards Association
- ASHVE—American Society of Heating and Ventilation
- ASTM—American Society for Testing Material
- BM—Bench mark
- CABRA—Copper and Brass Research Association
- CISPI—Cast Iron Soil Pipe Institute
- F & D—Faced and drilled
- IBBM—Iron body bronze or brass mounted
- MSS—Manufacturer Standardization Society of Valve and Fittings Industry
- NBS—National Bureau of Standards
- NDTS—Not drawn to scale
- NPS—Nominal pipe size
- OS & Y—Outside screw and yoke
- LIA—Lead Industries Association
- RNPT—Right-hand National Pipe Thread

Definitions

This section provides useful definitions of terms and quantities, plus facts and tips for plumbers, in the following areas:

- General information
- SI (metric system)
- Air pressure
- Absolute zero
- Cylinder pressure
- Welding flame
- Boiling points
- Melting points
- Minimum grade fall
- Gaskets

General Information

Following are general facts of interest:

- 1 pound of air pressure elevates water approximately 2.31 feet under atmospheric conditions of 14.72 psi.
- 2.3 feet of water equals 1 psi.
- 1 foot of water equals 0.434 psi.
- 1.728 cubic inches equals 1 cubic foot.
- 231 cubic inches equals 1 U.S. gallon.
- 1 cubic foot of water at 39°F weighs 62.48 lbs.
- 1 U.S. gallon of cold water weighs 8.33 lbs.
- 1 cubic foot of water contains 7.48 gallons.

SI (Metric System)

In SI (metric system), the following is true:

- 1 kilopascal (kPa) of air pressure elevates water ap-

proximately 10.2 cm under atmospheric conditions of 101 kPa.
- 10.2 cm of water equals 1 kPa.
- 51 cm of water equals 5 kPa.
- 1 meter of water equals 9.8 kPa.
- 10,000 square centimeters equals 1 square meter.
- 1 cubic meter equals 1,000,000 cubic centimeters (cm^3) or 1000 cubic decimeters (dm^3).
- 1 liter of cold water at 4°C weighs 1 kilogram.

Air Pressure

Following are some facts related to air pressure:

- One cubic inch of mercury weighs 0.49 lbs. Therefore, a 10-inch column of mercury would be 10 × 0.49 or 4.9 psi. Generally speaking, 2 inches of mercury is equivalent to 1 pound of pressure.
- One cubic foot of air weighs 1.2 ounces or 0.075 lbs.
- Atmospheric pressure of 14.7 psi will balance or support a column of mercury 29.92 inches high.

Absolute Zero

Following are some facts related to absolute zero:

- Absolute zero is –459.69°F.
- Absolute zero is –273.16°C.
- In SI, 1 cm of mercury at 0°C is equivalent to 1.3332239 kPa pressure. Therefore, a 24-cm column of mercury would be 24 × 1.3332239, or (rounded) 32 kPa. Generally speaking, 6 cm of mercury is equal to 8 kPa.
- One cubic meter of air weighs 1.214 kilograms.

Table 19-1 Boiling Points of Water at Various Pressures Above Atmospheric

Gage Pressure (psi or kPa)	Boiling Point
1–6.89	216°F (102.2°C)
4–27.58	225°F (107.2°C)
15–103.43	250°F (121.1°C)
25–172.36	267°F (130.5°C)
30–206.84	274°F (134.4°C)
45–310.26	293°F (145.0°C)
50–344.73	297°F (147.2°C)
65–448.13	312°F (155.5°C)
75–517.1	320°F (160.0°C)
90–620.52	335°F (168.3°C)
100–689.47	338°F (170.0°C)
125–861.83	353°F (178.3°C)
150–1034.2	366°F (185.5°C)

- Atmospheric pressure of 101.3 kPa will balance or support a column of mercury 76 cm high.

Cylinder Pressure
Cylinders are charged with oxygen at a pressure of 2000 psi (13,789 kPa) at 70°F (21°C).

Welding Flame
The temperature of an oxygen-acetylene flame is estimated to be more than 6000°F (3316°C).

Boiling Points
The boiling point at atmospheric pressure, or 0-gage pressure, is 212°F (100°C). Table 19-1 shows boiling points at various pressures above atmospheric.

Melting Points

Following are important melting points:

- Lead melts at 622°F (328°C).
- Tin melts at 449°F (231.78°C).
- 50-50 solder begins to melt at 362°F (183.3°C).
- Zinc melts at 790°F (421°C).
- Pure iron melts at 2730°F (1499°C).
- Steel melts at 2400°F–2700°F (1315.5°C–1482.2°C).

Minimum Grade Fall

The absolute minimum fall or grade for foundation or subsoil drainage lines is 1 inch (2.5 cm) in 20 feet (6.1 m).

Gaskets

Suitable gasket material for a flange union should be made of the following:

- *Cold-water piping*—Sheet rubber or asbestos sheet packing. (Asbestos is not permitted in any form in some local codes.)
- *Hot-water lines*—Rubber or asbestos composition.
- *Gas piping*—Leather or asbestos composition.
- *Oil lines*—Metallic or, where permitted, asbestos composition.
- *Gasoline conduction*—Metallic.

By applying graphite to one side of a gasket, removal at a later date is made much easier.

Screws and Bolts

This section provides reference information about screws and bolts, including the following:

- American Standard threads
- Machine screw bolt information
- Standard wood screw information

Table 19-2 American Standard Threads

Pipe Size	Threads per Inch	Threads per Centimeter
1/8 inch or 3.175 mm	27	10½
1/4 inch or 6.35 mm	18	7
3/8 inch or 9.525 mm	18	7
1/2 inch or 12.7 mm	14	5½
3/4 inch or 19.05 mm	14	5½
1 inch or 25.4 mm	11½	4½
1¼ inches or 31.75 mm	11½	4½
1½ inches or 38.1 mm	11½	4½
2 inches or 50.8 mm	11½	4½
2½ inches through 4 inches or 63.5 mm through 101.6 mm	8	3⅛

American Standard Threads
Table 19-2 shows standard thread measurements for common pipe sizes.

Machine Screw Bolt Information (NC)
Table 19-3 shows machine screw bolt information.

Standard Wood Screw Information
Table 19-4 shows standard wood screw information. Flathead screws are measured by overall length; roundhead screws from base of head to end.

Forming an Angle Using a Folding Rule
You can form commonly used angles by using the first four sections of a folding rule. This will enable you to determine what fittings will work in offsetting situations or the proper

Abbreviations, Definitions, and Symbols **381**

Table 19-3 Machine Screw Bolt Information

Size	Diameter in Inches or Millimeters to nearest $1/64$ inch ($1/2$ mm)
1	$5/64$ inch (2 mm)
2	$5/64$ inch (2 mm)
3	$7/64$ inch ($2 1/2$ mm)
4	$7/64$ inch (3 mm)
5	$1/8$ inch (3 mm)
6	$9/64$ inch ($3 1/2$ mm)
8	$11/64$ inch (4 mm)
10	$13/64$ inch (5 mm)
12	$13/64$ inch ($5 1/2$ mm)
$1/4$ inch	$1/4$ inch (6 mm)

Note: Above No. 12, machine screw sizes are designated by actual diameter.

Table 19-4 Wood Screw Information

Size Number	Decimal	Diameter Inches	Diameter mm*
0	0.064	$1/16$	$1 1/2$
1–2	0.077, 0.090	$3/32$	2
3–4–5	0.103, 0.116, 0.129	$1/8$	$2 1/2$–3
6–7–8	0.142, 0.155, 0.168	$5/32$	$3 1/2$–4
9–10–11–12	0.181, 0.194, 0.207, 0.220	$3/16$	$4 1/2$–$5 1/2$
14–16	0.246, 0.272	$1/4$	6–7
18–20	0.298, 0.324	$5/16$	$7 1/2$–8
24	0.376	$3/8$	$9 1/2$

*Rounded to nearest $1/2$ mm.

direction to turn a pipe when it is necessary to come out of a wall at an angle. Following are some examples of forming one of these angles or bends:

- $22\frac{1}{2}°$ or $\frac{1}{16}$ bend—Take the tip of the rule and touch $23\frac{3}{4}$ inches. Straighten out the rule at the second joint for the angle.
- $11\frac{1}{4}°$ or $\frac{1}{32}$ bend—Take the tip of the rule and touch $23\frac{15}{16}$ inches. Follow as previously mentioned.
- $30°$ angle—Touch the tip of the rule to $14\frac{13}{16}$ inches. Follow as before.
- $45°$ or $\frac{1}{8}$ bend—Touch the tip of the rule to 23 inches. Follow as before.
- $60°$ or $\frac{1}{6}$ bend—Touch $22\frac{1}{4}$ inches with the tip. Follow as before.
- $72°$ or $\frac{5}{8}$ bend—Touch the tip of the rule to $21\frac{5}{8}$ inches. Follow as before.
- $90°$ angle—Touch $20\frac{1}{4}$ inches with the tip. Follow as before.

Table 19-5 Hanger Rod Sizes

Iron Pipe Size	Rod Size
$\frac{1}{8}$ inch to $\frac{1}{2}$ inch (3–13 mm)	$\frac{1}{4}$ inch (6 mm)
$\frac{3}{4}$ inch to 2 inches (19–51 mm)	$\frac{3}{8}$ inch (10 mm)
$2\frac{1}{2}$ inches and 3 inches (64 cm and 76 mm)	$\frac{1}{2}$ inch (13 mm)
4 inches and 5 inches (10 cm and 13 cm)	$\frac{5}{8}$ inch (16 mm)
6 inches (15 cm)	$\frac{3}{4}$ inch (19 mm)
8 inches, 10 inches, 12 inches (20 cm, 25 cm, 30 cm)	$\frac{7}{8}$ inch (22 mm)
14 inches and 16 inches (36 cm and 41 cm)	1 inch (25 mm)

Abbreviations, Definitions, and Symbols **383**

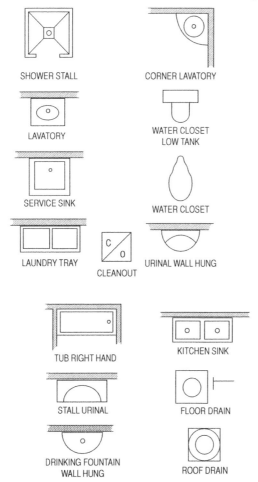

Fig. 19-1 Symbols for plumbing fixtures

Hanger Rod Sizes

Table 19-5 shows standard hanger rod sizes.

Hanger rod is threaded with national coarse (NC) bolt dies. A threading die marked "$5/8$ inch-11 NC" would be used for $5/8$-inch-diameter rod, the thread classified as national coarse with 11 threads per inch.

Symbols for Plumbing Fixtures

Fig. 19-1 shows common symbols used to denote plumbing fixtures.

20. FORMULAS

This chapter provides reference information about working with formulas, including the following:

- Working with squares and square roots
- Working with right triangles
- Determining capacities
- Working with areas
- Working with pressures
- Working with rolling offsets

Working with Squares and Square Roots

Table 20-1 shows the squares of common numbers.

Finding Square Roots

Pointing off the number to be worked in preparation to a solution is the first and most important step in finding square roots. Whether working with a decimal or whole number, always point off in twos from left to right. When a decimal is present and the numbers are odd, a zero (0) must be added, as shown here:

$\sqrt{25.325}$, should be $\sqrt{25.3250}$.

Following are some examples:

$\sqrt{2'59} \quad \sqrt{25.'32} \quad \sqrt{25.'32'50}$

Beginning with the first number or set of numbers on the left, we begin our problem. In the three examples just shown, the numbers would be 2, 25, and 25, respectively. We must find the nearest square (or number multiplied by itself) to each number which does not total higher than the number. In the examples just presented, squares would be 1, 5, and 5, respectively.

Chapter 20

In the next step, you will perform the following calculations:

```
        5           [1]  [4]  [8]
      × 5            5.  0   3
      ---         _____
       25        √25.'32'50'
                  [1]  25
      100        [3]  50 | 32  [2]
 [8]  + 3                3250  [5]
      ----              − 3009
      1003             --------
                         241  [7]
      1003
      ×  3
      ----
      3009
```

Following are the steps for performing the calculation (bracketed numbers in preceding equation correlate to steps):

1. $5 \times 5 = 25$ (place as shown).
2. Bring down the 32.
3. Try dividing 32 by 50 (the 5 + a trial 0).
4. 32 is less than 50, so place 0 above the 32.
5. Bring down the 50 that is alongside the 32 and place as shown.
6. Multiply the 50 above by 2 to obtain 100. Take this 100 and add a trial number (in this case, a 3) and multiply the 1003 by 3. The 3 is the number then used to make sure the 3250 is not exceeded when 3 is multiplied by 1003. If this exceeded the 3250, then you'd have to drop back and try a 2, and so on, until you get a result less than that desired.
7. Subtract 3009 from 3250 to obtain 241.

Table 20-1 Table of Squares

Number	Square
11	121
12	144
13	169
14	196
15	225
16	256
17	289
18	324
19	361
20	400
21	441
22	484
23	529
24	576
25	625
26	676

8. The 241 may be left as a remainder or you may continue to obtain 5.0323951. However, calculating two places beyond the decimal point is usually sufficient for this type of work.

This can be continued until the square root is found to be 5.0323951. However, today very inexpensive calculators save time and improve accuracy. Just enter 25.325 and then press the following button to instantly get the answer to seven decimal places:

$[\sqrt{}]$

388 Chapter 20

The following shows the problem proved:

```
    5.03
  × 5.03
  ------
    1509
   25150
  -------
  25.3009
   + 241
  -------
  25.3250 = R
```

Finding Diagonal of Square

The diagonal of a square equals the square root of the sum of the squares of the two sides (see Fig. 20-1).

As an example, find the diagonal of a square when the area is 8100 square inches ($90^2 = 8100$). Remember that area equals length × width.

[1] $8100 \times 2 = 16,200$

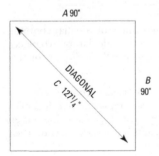

Fig. 20-1 A square produces two right angles when bisected or cut in half.

Formulas **389**

[2] $16,200 \text{ in}^2 =$ twice the area

[3] $\sqrt{16,200} = 127.27922$ or

[4] rounded to $127\frac{1}{4}''$ (see Fig. 19-1).

Note

Figures used in the example could also be centimeters or meters, and so on.

Working with Right Triangles

This section shows you how to work with right triangles. Fig. 20-2 shows a right triangle where the following is true:

- C is the hypotenuse.
- A is the altitude.
- B is the base.

Note

The square of the hypotenuse equals the sum of the squares of the other two sides (Pythagorean Theorem). Referring to Fig. 20-1, the formula would be $C^2 = A^2 + B^2$.

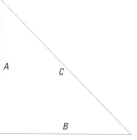

Fig. 20-2 Right triangle.

Note

To square your work while fabricating or forming a square, use the 3-4-5 method. You may use 3 feet, 4 feet, and 5 feet, or you may use 3 inches, 4 inches, and 5 inches. To increase the sides of the square, merely double or triple each number (see Fig. 20-3). For example, you would use 6-8-10, 9-12-15, and so on.

390 Chapter 20

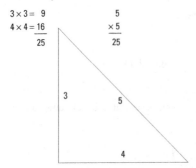

Fig. 20-3 Using the 3-4-5 method.

As a practical example of working with right triangles, imagine that you are running pipe or tubing in parallel runs using offsets and maintaining a uniform spread throughout the run, as shown in Fig. 20-4.

You have three choices:

- 22½° ells or ⅛ bends

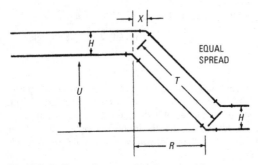

Fig. 20-4 Example runs of piping or tubing.

- 45° ells or ⅛ bends
- 60° ells or ⅙ bends

Thus, for the example shown in Fig. 20-4, you would use the following formulas:

$T = U \times 2.61$ $T = U \times 1.41$ $T = U \times 1.15$
or   or   or
$T = R \times 1.08$ $T = R \times 1.41$ $T = R \times 2$
$U = R \times 0.41$ $U = R$ $U = R \times 1.73$
$R = U \times 2.41$ $R = U$ $R = U \times 0.58$
$X = H \times 0.20$ $X = H \times 0.41$ $X = H \times 0.58$

Now, consider Fig. 20-5, where 45° ells are used.

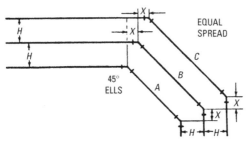

Fig. 20-5 Example using 45° ells.

For the example shown in Fig. 20-5, you would use the following formulas:

$X = H \times 0.41$
$B = A + (H \times 0.41 \times 2)$ or $B = C - (H \times 0.41 \times 2)$
$A = B - (H \times 0.41 \times 2)$
$C = B + (H \times 0.41 \times 2)$ or

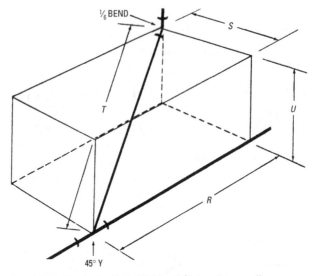

Fig. 20-6 Example using 45° Y and $\frac{1}{6}$ bend in an offset.

Finally, Fig. 20-6 shows a 45° Y and $\frac{1}{6}$ bend in an offset. For the example shown in Fig. 20-5, the following would be true:

$S = U$

$R = S \times 1.41$

$T = S \times 2, \text{or } U \times 2, \text{or } R \times 1.41$

$U = $ Vertical rise

$S = $ Horizontal spread

R = Advance or setback

T = Travel or pipe to be cut (will be center-to-center, or C-C, measurement)

Determining Capacities

This section shows how to determine tank capacities in gallons and in liters.

Determining Capacity of a Tank in Gallons

In this example, consider a scenario where the tank is 4 feet 0 inches in diameter and 10 feet 0 inches long:

1. Find the area of the circle (diameter squared × 0.7854):
   4 feet × 4 feet × 0.7854 = 12.566 square feet
2. Find the cubic contents (area of a circle × length):
   12.566 square feet × 10 feet = 125.66 cubic feet
3. Find the number of gallons by multiplying the cubic contents by 7.48 (the number of gallons in one cubic foot):
   125.66 cubic feet × 7.48 gallons = 939.936, or 940 gallons

Determining Capacity of a Tank in Liters

In this example, consider a scenario where the tank is 1.22 meters in diameter and 3.05 meters long:

1. Find the area of the circle (diameter squared × 0.7854):
   1.22 × 1.22 × 0.7854 = 1.169 square meters
2. Find the cubic contents (area of a circle × length):
   1.169 square meters × 3.05 meters = 3.565 cubic meters
3. Find the number of liters by multiplying the cubic contents by 1000 (the number of liters in one cubic meter):
   3.565 cubic meters × 1000 = 3565 liters

Working with Areas

To find the area of a triangle, multiply the base (B) times the height (H), and then multiply the product by $\frac{1}{2}$, as shown in the following formula:

$A = \frac{1}{2} BH$

To find the area of a circle, multiply the radius (R) by pi (π), or multiply the square of the diameter (D) by 0.7854, as shown in the following formulas:

$A = \pi R^2$

$A = 0.7854 D^2$

Note
$\pi = 3.141592654$

To find the area of a circle when the circumference is known, use the following formula:

$A = C^2 \over 12.57$

If a circle is 45^x (where x may be inches, feet, meters, centimeters, or millimeters), the calculation would be as follows:

$$A = \frac{45 \times 45}{12.56}$$

Answer: $A = 161.09^x$

Working with Pressures

To find the head (H) when pressure is given, divide the pressure by 0.433, or multiply the pressure by 2.309, as shown here:

Formulas 395

psi = $H \times 0.433$

H = psi $\times 2.309$

Working with Rolling Offsets

This section uses Fig. 20-7 as an example for calculating rolling offsets.

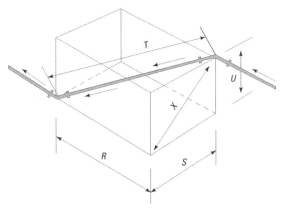

Fig. 20-7 Rolling offsets example.

Using Fig. 20-7 as a guide, use the following formula for a 45° fitting:

$X = \sqrt{S^2 + U^2}$

$T = X \times 1.41$

$R = X$

Using Fig. 20-7 as a guide, use the following formula for a 60° fitting:

$X = \sqrt{S^2 + U^2}$

$T = X \times 1.15$

$R = X \times 0.58$

Using Fig. 20-7 as a guide, use the following formula for a $22\frac{1}{2}°$ fitting:

$X = \sqrt{S^2 + U^2}$

$T = X \times 2.61$

$R = X \times 2.41$

As an example, using 45° fittings, assume U is 25 and S is 30; hence, U must be squared:

Step 1

$25 \times 25 = 625$ (U squared)

Step 2

S must be squared: or $30 \times 30 = 900$

Step 3

Add $U^2 =$ 625
$S^2 = +\ 900$
1525

Find the square root of 1525 (calculator solution):

$$\sqrt{1525.00}^{\,39.05}$$
$= $ sq. root R^4

Proved:

$$\begin{array}{r} 39.05 \\ \times\ 39.05 \\ \hline 1524.9025 \end{array}$$

So, in our example, the final step is as follows:

Step 4

$X = 39$

$T = $ Travel or pipe to be cut

$$\begin{array}{r} 1.41 \\ \times\ 39 \\ \hline 1269 \\ 423 \\ \hline 54.99 \end{array}$$

Answer: 55 C-C

Table 20-2 provides conversions from fractions to decimals. When multiplying, you may find it more convenient to change a fraction to a decimal.

Rolling Offset (Using 45° Fittings) Further Simplified

This section provides a simplified example of calculating rolling offset. In the example, the first offset is 6 inches and the second offset is 13 inches.

Table 20-2 Fraction-Decimal Equivalents

Fraction	Decimal
1/16 inch	0.06
1/8 inch	0.13
3/16 inch	0.19
1/4 inch	0.25
5/16 inch	0.31
3/8 inch	0.38
7/16 inch	0.44
1/2 inch	0.50
9/16 inch	0.56
5/8 inch	0.63
11/16 inch	0.69
3/4 inch	0.75
13/16 inch	0.81
7/8 inch	0.88
15/16 inch	0.94

Note
Square root and constant multiplication have been worked out in this example.

Following are the steps to completing this calculation:

1. Multiply each offset by itself.
2. Then, add both results together.
3. Refer to Table 20-3 later in the chapter to find the nearest number to the result you obtained.
4. Your problem will be completed center to center (C-C). After fitting takeoff, your cut should be to the nearest 1/16 inch or 1/32 inch.

Formulas **399**

Table 20-3 Results with Answers

Result	C-C Answer	Result	C-C Answer
13 inches	5^1/$_{16}$ inches	22.75 inches	6^3/$_4$ inches
13.25 inches	5^1/$_8$ inches	23.50 inches	6^{13}/$_{16}$ inches
13.50 inches	5^3/$_{16}$ inches	24 inches	6^7/$_8$ inches
13.75 inches	5^1/$_4$ inches	24.25 inches	6^{15}/$_{16}$ inches
14.25 inches	5^5/$_{16}$ inches	24.50 inches	7 inches
14.50 inches	5^3/$_8$ inches	25 inches	7^1/$_{16}$ inches
14.75 inches	5^7/$_{16}$ inches	25.50 inches	7^1/$_8$ inches
15.25 inches	5^1/$_2$ inches	26 inches	7^3/$_{16}$ inches
15.50 inches	5^9/$_{16}$ inches	26.25 inches	7^1/$_4$ inches
15.75 inches	5^5/$_8$ inches	26.75 inches	7^5/$_{16}$ inches
16.25 inches	5^{11}/$_{16}$ inches	27.25 inches	7^3/$_8$ inches
16.50 inches	5^3/$_4$ inches	27.75 inches	7^7/$_{16}$ inches
17 inches	5^{13}/$_{16}$ inches	28.25 inches	7^1/$_2$ inches
17.25 inches	5^7/$_8$ inches	28.75 inches	7^9/$_{16}$ inches
17.75 inches	5^{15}/$_{16}$ inches	29.25 inches	7^5/$_8$ inches
18 inches	6 inches	29.50 inches	7^{11}/$_{16}$ inches
18.50 inches	6^1/$_{16}$ inches	30 inches	7^3/$_4$ inches
18.75 inches	6^1/$_8$ inches	30.50 inches	7^{13}/$_{16}$ inches
19.25 inches	6^3/$_{16}$ inches	31.25 inches	7^7/$_8$ inches
19.75 inches	6^1/$_4$ inches	31.50 inches	7^{15}/$_{16}$ inches
20 inches	6^5/$_{16}$ inches	32 inches	8 inches
20.25 inches	6^3/$_8$ inches	32.50 inches	8^1/$_{16}$ inches
20.75 inches	6^7/$_{16}$ inches	33 inches	8^1/$_8$ inches
21.25 inches	6^1/$_2$ inches	33.50 inches	8^3/$_{16}$ inches
21.75 inches	6^9/$_{16}$ inches	34 inches	8^1/$_4$ inches
22 inches	6^5/$_8$ inches	34.75 inches	8^5/$_{16}$ inches
22.50 inches	6^{11}/$_{16}$ inches	35.25 inches	8^3/$_8$ inches

(*continued*)

Table 20-3 (*continued*)

Result	C-C Answer	Result	C-C Answer
35.75 inches	$8^{7}/_{16}$ inches	97.50 inches	$13^{15}/_{16}$ inches
37.25 inches	$8^{5}/_{8}$ inches	100 inches	$14^{1}/_{8}$ inches
40 inches	$8^{15}/_{16}$ inches	102.50 inches	$14^{1}/_{4}$ inches
41.25 inches	$9^{1}/_{16}$ inches	105 inches	$14^{7}/_{16}$ inches
42.50 inches	$9^{3}/_{16}$ inches	107.50 inches	$14^{5}/_{8}$ inches
45 inches	$9^{7}/_{16}$ inches	110 inches	$14^{3}/_{4}$ inches
45.25 inches	$9^{1}/_{2}$ inches	112.50 inches	$14^{15}/_{16}$ inches
47.50 inches	$9^{11}/_{16}$ inches	115 inches	$15^{1}/_{8}$ inches
50 inches	10 inches	117.50 inches	$15^{5}/_{16}$ inches
52.50 inches	$10^{3}/_{16}$ inches	120 inches	$15^{7}/_{16}$ inches
55 inches	$10^{7}/_{16}$ inches	122.50 inches	$15^{5}/_{8}$ inches
57.50 inches	$10^{11}/_{16}$ inches	125 inches	$15^{3}/_{4}$ inches
60 inches	$10^{15}/_{16}$ inches	127.50 inches	$15^{15}/_{16}$ inches
62.50 inches	$11^{1}/_{8}$ inches	130 inches	$16^{1}/_{16}$ inches
65 inches	$11^{3}/_{8}$ inches	132.50 inches	$16^{3}/_{16}$ inches
67.50 inches	$11^{7}/_{16}$ inches	135 inches	$16^{3}/_{8}$ inches
70 inches	$11^{13}/_{16}$ inches	137.50 inches	$16^{1}/_{2}$ inches
72.50 inches	12 inches	140 inches	$16^{11}/_{16}$ inches
75 inches	$12^{3}/_{16}$ inches	142.50 inches	$16^{13}/_{16}$ inches
77.50 inches	$12^{7}/_{16}$ inches	145 inches	17 inches
80 inches	$12^{5}/_{8}$ inches	147.50 inches	$17^{1}/_{8}$ inches
82.50 inches	$12^{13}/_{16}$ inches	150 inches	$17^{1}/_{4}$ inches
85 inches	13 inches	152.50 inches	$17^{3}/_{8}$ inches
87.50 inches	$13^{3}/_{16}$ inches	155 inches	$17^{7}/_{16}$ inches
90 inches	$13^{3}/_{8}$ inches	157.50 inches	$17^{11}/_{16}$ inches
92.50 inches	$13^{7}/_{16}$ inches	160 inches	$17^{13}/_{16}$ inches
95 inches	$13^{3}/_{4}$ inches	162.50 inches	$17^{15}/_{16}$ inches

(*continued*)

Formulas **401**

Table 20-3 (*continued*)

Result	C-C Answer	Result	C-C Answer
165 inches	18$1/8$ inches	232.50 inches	21$7/16$ inches
167.50 inches	18$1/4$ inches	235 inches	21$5/8$ inches
170 inches	18$3/8$ inches	237.50 inches	21$3/4$ inches
172.50 inches	18$1/2$ inches	240 inches	21$13/16$ inches
175 inches	18$5/8$ inches	242.50 inches	21$15/16$ inches
177.50 inches	18$3/4$ inches	245 inches	22$1/16$ inches
180 inches	18$15/16$ inches	247.50 inches	22$3/16$ inches
182.50 inches	19$1/16$ inches	250 inches	22$5/16$ inches
185 inches	19$3/16$ inches	252.50 inches	22$7/16$ inches
187.50 inches	19$5/16$ inches	255 inches	22$1/2$ inches
190 inches	19$7/16$ inches	257.50 inches	22$5/8$ inches
192.50 inches	19$9/16$ inches	260 inches	22$3/4$ inches
195 inches	19$11/16$ inches	262.50 inches	22$13/16$ inches
197.50 inches	19$13/16$ inches	265 inches	22$15/16$ inches
200 inches	19$15/16$ inches	267.50 inches	23$1/16$ inches
202.50 inches	20$1/16$ inches	270 inches	23$3/16$ inches
205 inches	20$3/16$ inches	272.50 inches	23$1/4$ inches
207.50 inches	20$5/16$ inches	275 inches	23$3/8$ inches
210 inches	20$7/16$ inches	277.50 inches	23$1/2$ inches
212.50 inches	20$9/16$ inches	280 inches	23$7/16$ inches
215 inches	20$11/16$ inches	282.50 inches	23$11/16$ inches
217.50 inches	20$13/16$ inches	285 inches	23$13/16$ inches
220 inches	20$15/16$ inches	287.50 inches	23$7/8$ inches
222.50 inches	21 inches	290 inches	24 inches
225 inches	21$1/8$ inches	292.50 inches	24$1/8$ inches
227.50 inches	21$1/4$ inches	295 inches	24$3/16$ inches
230 inches	21$3/8$ inches	297.50 inches	24$5/16$ inches

(*continued*)

Chapter 20

Table 20-3 *(continued)*

Result	C-C Answer	Result	C-C Answer
300 inches	24⁷/₁₆ inches	367.50 inches	27 inches
302.50 inches	24¹/₂ inches	370 inches	27¹/₈ inches
305 inches	24⁵/₈ inches	372.50 inches	27³/₁₆ inches
307.50 inches	24³/₄ inches	375 inches	27⁵/₁₆ inches
310 inches	24¹³/₁₆ inches	377.50 inches	27³/₈ inches
312.50 inches	24¹⁵/₁₆ inches	380 inches	27¹/₂ inches
315 inches	25 inches	382.50 inches	27⁷/₁₆ inches
317.50 inches	25¹/₈ inches	385 inches	27¹¹/₁₆ inches
320 inches	25¹/₄ inches	387.50 inches	27³/₄ inches
322.50 inches	25⁵/₁₆ inches	390 inches	27¹³/₁₆ inches
325 inches	25³/₈ inches	392.50 inches	27¹⁵/₁₆ inches
327.50 inches	25¹/₂ inches	395 inches	28 inches
330 inches	25⁵/₈ inches	397.50 inches	28¹/₈ inches
332.50 inches	25¹¹/₁₆ inches	400 inches	28³/₁₆ inches
335 inches	25¹³/₁₆ inches	402.50 inches	28⁵/₁₆ inches
337.50 inches	25⁷/₈ inches	405 inches	28³/₈ inches
340 inches	26 inches	407.50 inches	28⁷/₁₆ inches
342.50 inches	26¹/₁₆ inches	410 inches	28⁹/₁₆ inches
345 inches	26³/₁₆ inches	412.50 inches	28⁵/₈ inches
347.50 inches	26⁵/₁₆ inches	415 inches	28¹¹/₁₆ inches
350 inches	26³/₈ inches	417.50 inches	28¹³/₁₆ inches
352.50 inches	26⁷/₁₆ inches	420 inches	28⁷/₈ inches
355 inches	26⁹/₁₆ inches	422.50 inches	29 inches
357.50 inches	26⁵/₈ inches	425 inches	29¹/₁₆ inches
360 inches	26³/₄ inches	427.50 inches	29¹/₈ inches
362.50 inches	26⁷/₈ inches	430 inches	29¹/₄ inches
365 inches	26¹⁵/₁₆ inches	432.50 inches	29⁵/₁₆ inches

(continued)

Table 20-3 (continued)

Result	C-C Answer	Result	C-C Answer
435 inches	$29^3/_8$ inches	502.50 inches	$31^5/_8$ inches
437.50 inches	$29^1/_2$ inches	505 inches	$31^{11}/_{16}$ inches
440 inches	$29^9/_{16}$ inches	507.50 inches	$31^3/_4$ inches
442.50 inches	$29^5/_8$ inches	510 inches	$31^{13}/_{16}$ inches
445 inches	$29^3/_4$ inches	512.50 inches	$31^{15}/_{16}$ inches
447.50 inches	$29^{13}/_{16}$ inches	515 inches	32 inches
450 inches	$29^7/_8$ inches	517.50 inches	$32^1/_{16}$ inches
452.50 inches	30 inches	520 inches	$32^1/_8$ inches
455 inches	$30^1/_{16}$ inches	522.50 inches	$32^1/_4$ inches
457.50 inches	$30^3/_{16}$ inches	525 inches	$32^5/_{16}$ inches
460 inches	$30^1/_4$ inches	527.50 inches	$32^3/_8$ inches
462.50 inches	$30^5/_{16}$ inches	530 inches	$32^7/_{16}$ inches
465 inches	$30^3/_8$ inches	532.50 inches	$32^9/_{16}$ inches
467.50 inches	$30^1/_2$ inches	535 inches	$32^5/_8$ inches
470 inches	$30^7/_{16}$ inches	537.50 inches	$32^{11}/_{16}$ inches
472.50 inches	$30^5/_8$ inches	540 inches	$32^3/_4$ inches
475 inches	$30^3/_4$ inches	542.50 inches	$32^{13}/_{16}$ inches
477.50 inches	$30^{13}/_{16}$ inches	545 inches	$32^{15}/_{16}$ inches
480 inches	$30^7/_8$ inches	547.50 inches	33 inches
482.50 inches	$30^{15}/_{16}$ inches	550 inches	$33^1/_{16}$ inches
485 inches	$31^1/_8$ inches	552.50 inches	$33^1/_8$ inches
487.50 inches	$31^1/_8$ inches	555 inches	$33^3/_{16}$ inches
490 inches	$31^3/_{16}$ inches	557.50 inches	$33^5/_{16}$ inches
492.50 inches	$31^5/_{16}$ inches	560 inches	$33^3/_8$ inches
495 inches	$31^3/_8$ inches	562.50 inches	$33^7/_{16}$ inches
497.50 inches	$31^7/_{16}$ inches	565 inches	$33^1/_2$ inches
500 inches	$31^1/_2$ inches	567.50 inches	$33^7/_{16}$ inches

(continued)

Table 20-3 *(continued)*

Result	C-C Answer	Result	C-C Answer
570 inches	33$^{11}/_{16}$ inches	637.50 inches	35$^{5}/_{8}$ inches
572.50 inches	33$^{3}/_{4}$ inches	640 inches	35$^{11}/_{16}$ inches
575 inches	33$^{13}/_{16}$ inches	642.50 inches	35$^{3}/_{4}$ inches
577.50 inches	33$^{7}/_{8}$ inches	645 inches	35$^{13}/_{16}$ inches
580 inches	33$^{15}/_{16}$ inches	647.50 inches	35$^{7}/_{8}$ inches
582.50 inches	34 inches	650 inches	35$^{15}/_{16}$ inches
585 inches	34$^{1}/_{8}$ inches	652.50 inches	36 inches
587.50 inches	34$^{3}/_{16}$ inches	655 inches	36$^{1}/_{16}$ inches
590 inches	34$^{1}/_{4}$ inches	657.50 inches	36$^{1}/_{8}$ inches
592.50 inches	34$^{5}/_{16}$ inches	660 inches	36$^{1}/_{4}$ inches
595 inches	34$^{3}/_{8}$ inches	662.50 inches	36$^{5}/_{16}$ inches
597.50 inches	34$^{7}/_{16}$ inches	665 inches	36$^{3}/_{8}$ inches
600 inches	34$^{9}/_{16}$ inches	667.50 inches	36$^{7}/_{16}$ inches
602.50 inches	34$^{5}/_{8}$ inches	670 inches	36$^{1}/_{2}$ inches
605 inches	34$^{11}/_{16}$ inches	672.50 inches	36$^{7}/_{16}$ inches
607.50 inches	34$^{3}/_{4}$ inches	675 inches	36$^{5}/_{8}$ inches
610 inches	34$^{13}/_{16}$ inches	677.50 inches	36$^{11}/_{16}$ inches
612.50 inches	34$^{7}/_{8}$ inches	680 inches	36$^{3}/_{4}$ inches
615 inches	34$^{15}/_{16}$ inches	682.50 inches	36$^{13}/_{16}$ inches
617.50 inches	35$^{1}/_{16}$ inches	685 inches	36$^{7}/_{8}$ inches
620 inches	35$^{1}/_{8}$ inches	687.50 inches	37 inches
622.50 inches	35$^{3}/_{16}$ inches	690 inches	37$^{1}/_{16}$ inches
625 inches	35$^{1}/_{4}$ inches	692.50 inches	37$^{1}/_{8}$ inches
627.50 inches	35$^{5}/_{16}$ inches	695 inches	37$^{3}/_{16}$ inches
630 inches	35$^{3}/_{8}$ inches	697.50 inches	37$^{1}/_{4}$ inches
632.50 inches	35$^{7}/_{16}$ inches	700 inches	37$^{5}/_{16}$ inches
635 inches	35$^{1}/_{2}$ inches	702.50 inches	37$^{3}/_{8}$ inches

(continued)

Table 20-3 (*continued*)

Result	C-C Answer	Result	C-C Answer
705 inches	$37^7/_{16}$ inches	757.50 inches	$38^{13}/_{16}$ inches
707.50 inches	$37^1/_2$ inches	760 inches	$38^7/_8$ inches
710 inches	$37^7/_{16}$ inches	762.50 inches	$38^{15}/_{16}$ inches
712.50 inches	$37^5/_8$ inches	765 inches	39 inches
715 inches	$37^{11}/_{16}$ inches	767.50 inches	$39^1/_{16}$ inches
717.50 inches	$37^3/_4$ inches	770 inches	$39^1/_8$ inches
720 inches	$37^{13}/_{16}$ inches	772.50 inches	$39^3/_{16}$ inches
722.50 inches	$37^7/_8$ inches	775 inches	$39^1/_4$ inches
725 inches	$37^{15}/_{16}$ inches	777.50 inches	$39^5/_{16}$ inches
727.50 inches	38 inches	780 inches	$39^3/_8$ inches
730 inches	$38^1/_8$ inches	782.50 inches	$39^7/_{16}$ inches
732.50 inches	$38^3/_{16}$ inches	785 inches	$39^1/_2$ inches
735 inches	$38^1/_4$ inches	787.50 inches	$39^9/_{16}$ inches
737.50 inches	$38^5/_{16}$ inches	790 inches	$39^5/_8$ inches
740 inches	$38^3/_8$ inches	792.50 inches	$39^{11}/_{16}$ inches
742.50 inches	$38^7/_{16}$ inches	795 inches	$39^3/_4$ inches
745 inches	$38^1/_2$ inches	797.50 inches	$39^{13}/_{16}$ inches
747.50 inches	$38^7/_{16}$ inches	800 inches	$39^7/_8$ inches
752.50 inches	$38^{11}/_{16}$ inches	802.50 inches	$39^{15}/_{16}$ inches
755 inches	$38^3/_4$ inches	805 inches	40 inches

Following is the calculation:

1st Offset	2nd Offset	
6″	13″	169″
×6″	×13″	+36″
36″	39″	205″ = RESULT
	+13	
	169″	

Rolling Offset Further Simplified

Table 20-3 provides 365 answers that eliminate square root.

Note

By looking at Table 20-3, you will note your answer to be 20 $^{3}/_{16}$ inches C-C using 45° fittings.

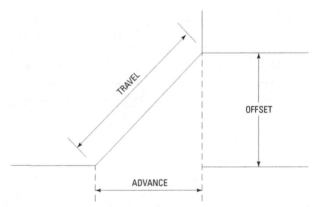

Fig. 20-8 Constants for calculating offset measurements.

Table 20-4 Constants for Calculating Offset Measurements

Degree of Fitting	Known Constant Factor
$5\frac{5}{8}°$	Offset × 10.20 = Travel
$5\frac{5}{8}°$	Offset × 10.152 = Advance
$11\frac{1}{4}°$	Offset × 5.126 = Travel
$11\frac{1}{4}°$	Offset × 5.027 = Advance
$22\frac{1}{2}°$	Offset × 2.613 = Travel
$22\frac{1}{2}°$	Offset × 2.414 = Advance
30°	Offset × 2.000 = Travel
30°	Offset × 1.732 = Advance
45°	Offset × 1.414 = Travel
45°	Offset × 1.00 = Advance
60°	Offset × 1.155 = Travel
60°	Offset × 0.577 = Advance
$67\frac{1}{2}°$	Offset × 1.083 = Travel
$67\frac{1}{2}°$	Offset × 0.414 = Advance
72°	Offset × 1.052 = Travel
72°	Offset × 0.325 = Advance

Constants for Calculating Offset Measurements

Fig. 20-8 shows constants for calculating offset measurements, and Table 20-4 shows how these relate to certain degrees of fittings.

Note

When these constants are used to calculate an offset measurement, the given offset and solution are expressed in center-to-center (C-C) measurements.

21. METRIC INFORMATION HELPFUL TO THE PIPING INDUSTRY

This chapter discusses the use of the metric system in the plumbing industry by reviewing the following topic areas:
- Metric abbreviations
- Metric conversions
- Pressure
- Blueprint scales

The building trades and the architectural community seldom use the term centimeter (cm) any longer, but rather commonly use millimeters (mm) for the smaller measurements. Usually the dimension will be given in inches, followed by the metric equivalent in parentheses.

As you may recall, England and Canada now use the metric system of measurement. However, to make this book more useful, both the system used in the U.S. and the metric system are referenced here.

Occasionally, you may need to convert one to the other. The formulas and equivalents are presented in this chapter for your reference.

Metric Abbreviations
Table 21-1 shows common metric abbreviations.

Metric Conversions
This section provides several tables to aid in converting between the metric and avoirdupois (U.S. or English) systems of weights and measurements.

Table 21-2 provides conversions for units of liquid volume.
Table 21-3 provides conversions for units of length.
Table 21-4 provides conversions for inches to millimeters.

Table 21-1 Metric Abbreviations

Abbreviation	Meaning
ANMC	American National Metric Council
ANSI	American National Standards Institute
cm	centimeter
cm^2	square centimeter
cm^3	cubic centimeter
dm	decimeter
dm^2	square decimeter
dm^3	cubic decimeter
g	gram
inHg	inches of mercury
J	joule
kg	kilogram
km	kilometer
km^2	square kilometer
kPa	kilopascal
L	liter
m	meter
m^2	square meter
mm^3	cubic millimeter
mg	milligram
mL	milliliter
mm	millimeter
NBS	National Bureau of Standards
N-m	newton-meter
°C	degrees Celsius
Pa	pascal
SI	International System of Units

Metric Information Helpful to the Piping Industry 411

Table 21-2 Liquid Volume

Metric	English
3.7854 L	1 gallon
0.946 L	1 quart
0.473 L	1 pint
1 L	0.264 gallons or 1.05668 quarts
1 L	33.814 ounces
29.576 mm	1 fluid ounce
236.584 mm	1 cup

Note: 1 liter contains 1000 milliliters.

Table 21-3 Length

Metric	English
1 m	39.37 inches
1 m	1000 millimeters
1 m	100 centimeters
1 m	10 decimeters
1 km	0.625 miles
1.609 km	1 mile
25.4 mm	1 inch
2.54 cm	1 inch
304.8 mm	1 foot
1 mm	0.03937 inch
1 cm	0.3937 inch
1 dm	3.937 inches

Table 21-4 Inches to Millimeters

Inches to Millimeters	
English (inches)	Metric (mm)
1	25.4
2	50.8
3	76.2
4	101.6
5	127
6	152.4
7	177.8
8	203.2
9	228.6
10	254
11	279.4
12	304.8

Note: Round off to nearest millimeter. For example,
4 inches = 102 mm
3 inches = 76 mm
2 inches = 51 mm

Table 21-5 provides conversions for parts of an inch to millimeters.

Table 21-6 provides conversions for inches and parts of an inch to centimeters.

Table 21-7 provides conversions for feet to millimeters.

Table 21-8 provides conversions for inches and feet to meters.

Table 21-9 provides conversions for units of weight (mass).

Table 21-10 provides conversions for squares and cube measurements.

Table 21-5 Parts of an Inch to Millimeters

English (Parts of an Inch)	Metric (mm)	English (Parts of an Inch)	Metric (mm)
$1/32$	0.79375 (0.80)	$9/16$	14.2875 (14.3)
$1/16$	1.5875 (1.6)	$5/8$	15.8750 (15.9)
$1/8$	3.175 (3.2)	$11/16$	17.4625 (17.5)
$3/16$	4.7625 (4.8)	$3/4$	19.0500 (19.1)
$1/4$	6.35 (6.4)	$13/16$	20.6375 (20.6)
$5/16$	7.9375 (7.9)	$7/8$	22.2250 (22.2)
$3/8$	0.5250 (9.5)	$15/16$	23.8175 (23.8)
$7/16$	11.1125 (11.1)	1	25.4000 (25.4)
$1/2$	12.7000 (12.7)		

Note: In most cases it is best to round off to the nearest millimeter. For example,
17.4625 would be 17 mm.
20.6375 would be 21 mm.

Table 21-6 Inches and Parts of an Inch to Centimeters

Inches (Parts of an Inch)	Metric (cm)	Inches (Parts of an Inch)	Metric (cm)
1/16	0.15875	2½	6.35
1/8	0.3175	3	7.62
1/4	0.635	4	10.16
3/8	0.9525	5	12.70
1/2	1.27	6	15.24
5/8	1.5875	7	17.78
3/4	1.905	8	20.32
7/8	2.2225	9	22.86
1	2.54	10	25.40
1¼	3.175	11	27.94
1½	3.81	12	30.48
2	5.08		

Table 21-7 Feet to Millimeters

English (feet)	Metric (mm)
2	609.6
3	914.4
4	1219.2
5	1524.0
6	1828.8
7	2133.6
8	2438.4
9	2743.2
10	3048.0
20	6096.0

Table 21-8 Inches and Feet to Meters

Inches	Meters	Feet	Meters
1	0.0254	$1\frac{1}{4}$	0.381
2	0.0508	$1\frac{1}{2}$	0.4572
3	0.0762	2	0.6096
4	0.1016	$2\frac{1}{2}$	0.762
5	0.127	3	0.9144
6	0.1524	4	1.2192
7	0.1778	5	1.524
8	0.2032	6	1.8288
9	0.2286	10	3.048
10	0.254	25	7.62
11	0.2794	50	15.24
12	0.3048	100	30.48

Table 21-9 Weight (Mass)

Measurement	Equivalent
1 kg	2.204623 pounds
453.592 g	1 kg
1 g	0.035 ounce
28.349 g	1 ounce
28,349 mg	1 ounce
1 g	1000 mg
1 kg	1,000,000 mg
1 kg	1000 g
0.02831 kg	1 ounce
1 lb	453,592.37 mg
1 lb	453.59237 g
1 lb	0.453592 kg
1 metric ton	1000 kg
1 metric ton	2204.623 pounds
1 mL water	1 g
1 L cold water (40°C)	1 kg

Note

Units of pressure, including the kilopascal (kPa), are discussed later in this chapter in the section *Pressure*.

Table 21-11 provides conversions between the Fahrenheit and Celsius temperature systems.

The following formulas may be used for converting temperatures given on one scale to that of the other. (This is easier to compute with a calculator.)

$$F = 1.8 \times C + 32$$
$$C = (F - 32) \div 1.8$$

Table 21-10 Square and Cube Measurements

Measurement	Equivalent
2.59 km^2	1 square mile
0.093 m^2	1 square foot
6.451 cm^2	1 square inch
0.765 m^3	1 cubic yard
0.028316 m^3	1 cubic foot
16.387 cm^3	1 cubic inch
1 m^3	35.3146 cubic feet
929.03 cm^2	1 square foot
10,000 cm^2	1 m^2
1 m^3	1,000,000 cm^3 or 1000 dm^3
10.2 cm of water	1 kPa of pressure
51 cm of water	5 kPa
1 m of water	9.8 kPa
1 m^3 of air	1.214 kg
1 cubic foot	28,316.846522 cm^3

Note

When designating temperatures as either Fahrenheit (F) or Celsius (C), the common practice is not to include a space between the degree symbol (°) and the letter designating the temperature scale (for example, 20°C).

Another measurement of temperature is the British thermal unit (Btu). A common conversion is between Btu and joules (the amount of work done by a force of one newton acting through a distance of one meter). Table 21-12 shows common conversions between Btu and joules.

Table 21-11 Temperature

Fahrenheit		Celsius
212°	Temperature of boiling water	100°
176°		80°
140°		60°
122°		50°
104°		40°
98.6°	Temperature of human body	37°
95°		35°
86°		30°
77°		25°
68°		20°
50°		10°
32°	Temperature of melting ice	0°
–4°		–20°
–40°	Temperature equal	–40°
–459.69°	Absolute zero	–273.16°

Table 21-12 Btu and Joules

Btu	Joule
1.0 (mean)	1.05587
1.0 (international)	1.055056
144 (amount required to change 1 pound of ice to water)	152

Metric Information Helpful to the Piping Industry **419**

Table 21-13 Pressure Conversions

Measurement	Equivalent
1 pound per square inch (psi)	6.894757 kPa
1 m column of water	9.794 kPa (or 0.2476985 kPa per inch)
1 cm column of mercury at 0°C	1.3332239 kPa
10.2 cm of water	1 kPa
51 cm of water	5 kPa
1 inch of mercury (inHg)	3.386389 kPa
6 cm of mercury	8 kPa

Pressure

The kilopascal (kPa) is the unit of measurement recommended for fluid pressure for almost all fields of use, such as barometric pressure, gas pressure, tire pressure, and water pressure.

Table 21-13 shows common pressure conversions.

When working with pressure, remember that atmospheric pressure is commonly measured at 101.3 kPa metric and 14.7 psi English. A common example of plumbers working with pressures is determining pressures for columns of water. When performing pressure calculations for columns of water, keep in mind the following:

- Atmospheric pressure of 101.3 kPa will balance or support a column of mercury 76 cm high.
- To find head pressure in decimeters when pressure is given in kPa, divide pressure by 0.9794.
- To find head pressure in kPa of a column of water given in decimeters, multiply the number of decimeters by 0.9794.

420 Chapter 21

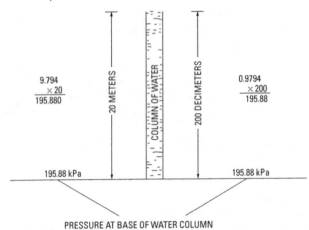

Fig. 21-1 Pressure example: 1 kPa of air pressure elevates water approximately 10.2 cm under atmospheric conditions of 101 kPa.

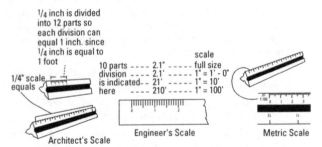

Fig. 21-2 Ways of measuring: an architect's scale, an engineer's scale, and a metric scale.

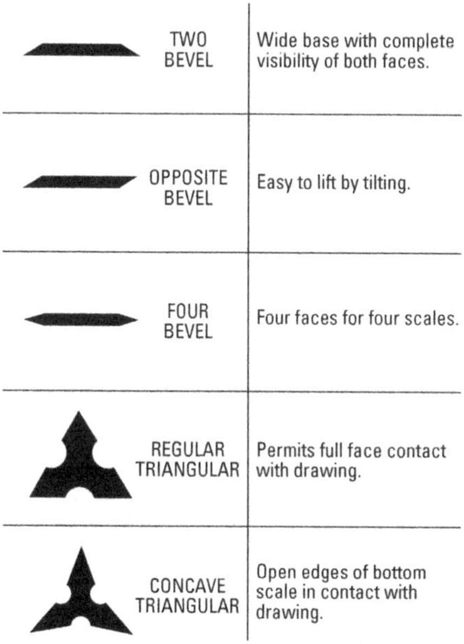

	TWO BEVEL	Wide base with complete visibility of both faces.
	OPPOSITE BEVEL	Easy to lift by tilting.
	FOUR BEVEL	Four faces for four scales.
	REGULAR TRIANGULAR	Permits full face contact with drawing.
	CONCAVE TRIANGULAR	Open edges of bottom scale in contact with drawing.

Fig. 21-3 End views of scales used by drafters.

422 Chapter 21

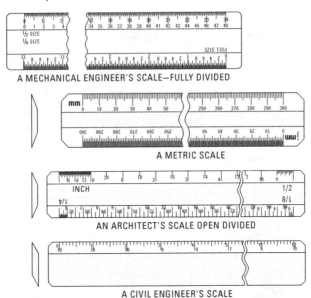

Fig. 21-4 Four types of scales used by drafters.

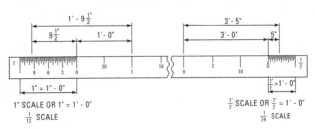

Fig. 21-5 Architect's scale.

- To find head pressure in meters when pressure is given in kPa, divide the number of meters by 9.794.
- To find pressure in kPa of a column of water given in meters, multiply meters by 9.794 (see Fig. 21-1).

Blueprint Scales

When reading blueprints, it is best to know that there are three types of scales used to make drawings for buildings: the *architect's scale*, the *engineer's scale*, and the *metric scale* (see Fig. 21-2). Note the end views of the scales that Frederick Post Company makes available for use in residential, commercial, and industrial drafting rooms, as shown in Fig. 21-3.

Fig. 21-4 shows the mechanical engineer's scale, the metric scale, the architect's scale, and the civil engineer's scale. The mechanical engineer's scale is usually triangular, as shown in Fig. 21-4. The civil engineer's scale is divided into 10 marks per inch, so that means each division is one-tenth (0.1) of an inch. The architect's scale has the inch divided into 16 parts, so each represents $1/16$ inch, as shown in Fig. 21-5.

22. KNOTS COMMONLY USED

In some instances, plumbers are required to know how to make knots to slip around pipes (especially large concrete pipes and heavy steel pipes) so that they can be lifted and placed in trenches or properly stacked for storage purposes.

Fig. 22-1 shows some common knots.

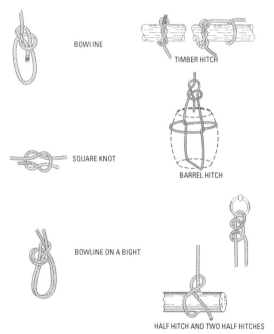

Fig. 22-1 Common knots.

23. TYPICAL HOISTING SIGNALS

In some instances, plumbers find themselves in situations where verbal communication is not possible. This may be particularly true when loading or unloading materials (especially large concrete pipes and heavy steel pipes) at a job site when noise or location prevent the plumber from verbally communicating with a truck driver or heavy equipment operator. In these situations, hand signals such as those shown in Fig. 23-1 may be required.

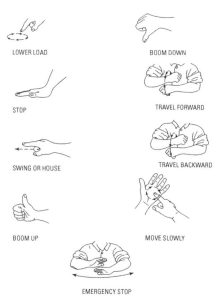

Fig. 23-1 Typical hoisting signals.

Index

A

abbreviations, used by plumbers, 375
absolute pressure, 7
absolute zero, 377
ABS pipe. *See* acrylonitrile-butadiene-styrene (ABS) pipe
acetylene cylinders, 180
acid-diluting tanks, 52
acrylonitrile-butadiene-styrene (ABS) pipe
  absorption of heat, 196
  allowing for expansion and contraction in, 203
  connecting to other materials, 197, 200–201
  description of, 183
  for multistory stacks, 201–202
  storage of, 194–195
  water pressure ratings for, 194
Act-O-Matic shower heads
  adjusting direction of, 368
  features of, 361
  figure of, 366
  installation of, 366–368
  measurements for, 370
  parts list for, 369
  pre-installation considerations for, 361
  repair kit for, 371
  water lines for, 361
adapters
  for copper pipe, 287
  CPVC-to-metal, 198–199
  for drainage systems, 201
  See also unions
adjustable bracket, 112
adjustable tailpiece, 347–348
adjustable wrench, 22
adjusting nut, 111
aerators
  on Aquarian sink fitting, 82
  at base of stack, 301
  diagram of, 298
  functions of, 296, 297
  parts of, 296–297
  for soil branch, 301
air chambers, 123
  for air separation, 298
  in diagram of home plumbing system, 212
air compressors
  air piping for, 37
  for dental offices, 33, 35
air gaps, 9
air induction controls, 104
air pressure, 377
  calculating, 394–395
  necessary for elevating water, 420
  needed to elevate water, 376
air separation chamber, 298
air supply
  to dental chairs, 37
  for dental offices, 41
American Standard
  Aquaseal valve, 80, 81
  thread measurements, 380
Americans with Disabilities Act (ADA), guidelines for sink installation, 62–63

Index

annealed tube, 279, 280–281
antisiphon hole, 13
Aquarian sink fitting, 82
Aquaseal sink fitting, 77–78
Aquaseal valve, 80, 81
architect's scale, 420–422, 423
ARC pipe, 279, 281
area, calculating, 394
atmospheric pressure, 377, 419, 423
augers, 25

B

back vent, 129
baffle, 297
Bak-Check control stop, 345, 346, 347
ball bearing thrust collar, 111
ball cocks
  figure of, 64
  replacing, 65–66
ballpeen hammer, 19, 157
ball seals, 61
ball thrust bearing, 114
barrel hitch knot, 426
base elevation, 3
basement floor level, 3
basin inlet, 107
basin wrench, 23
bathrooms
  fixture connections in, 58
  lavatories, 59
  location of fixtures in, 57
  sewer and vent system for, 58
  short history of, 15
  water closets in, 59
bathtubs
  approximate water demand for, 285
  checking for leaks in, 92
  cleaning, 101
  combined with shower, 87, 88
  corner, 92, 94
  installing overflow in, 67–69
  installing trip waste in, 67–69
  minimum copper tube sizes for short-branch connections to, 283
  oval, 92, 93
  rough-in of, 66, 67
  symbol for, 383
  venting system for, 137, 141
  whirlpool (*see* whirlpool bathtubs)
batteries, replacing in flushometer, 359–360
bell-and-spigot joint, 160
bell-and-spigot soil pipe, 161
bell end pipe, putting joints in, 185–190
bench mark
  abbreviation for, 375
  defined, 3
bends
  making, 382
  minimum offsets for, 273, 274, 275–276
beveling, 146–147
Bibb screws, removing, 72
blowout urinals, 340
blueprints
  reading, 3–4
  scales for, 420–422, 423
boiler drain, 125
boiling points, 378
bolts
  reference information for, 379–380, 381
  threading machine for, 22
boom down, 428
boom up, 428
bowline knot, 426
branch inlet, 297
branch vent, 129
brass fittings, 9
brass pipe, expansion of, 123
brazing. *See* silver brazing
brazing alloy, 167, 169

Index **431**

British thermal unit (Btu), conversion to joules, 417, 418
building drain, 10
building sewer, 10
burner flange, 318
burners
 figure of, 13
 installation diagram, 318
 lighting, 320–322
 for water heaters, 318–319
burrs, removing, 173, 186
bushing, 163
butt joints, 153
bypass orifice, 363

C

cap, 163
capacities
 of tanks in gallons, 393
 of tanks in liters, 393
carbon tetrachloride (CCl_4) extinguisher, 16–17
cartridge seal set, 82
casing bearing, 115
cast-iron no-hub pipe fittings
 combination Y and $1/4$ bends, 253, 255–257
 increaser-reducer, 273, 275
 list of, 252–253
 minimum offsets for bends in, 273, 274, 275–276
 $1/2$ P-traps, 273, 274
 $1/2$ S-traps, 273, 274
 $1/4$ bends, 261–262, 266, 269–270, 271–272
 1/5 bends, 258
 1/6 bends, 258
 1/8 bends, 258, 259
 1/16 bends, 258, 259
 sanitary T-branches, 262, 263–266
 sweeps, 259–261
 tapped extension place, 273, 274–275
 test tees, 261, 262, 263
 Y-branches, 253, 254, 255, 266, 267–269
cast-iron no-hub systems
 advantages of, 235, 236, 237
 clamping pipe at floor in, 250
 horizontal piping in, 241, 243–245
 installation of, 238–240
 overview of, 238
 pipe through floor slabs in, 246
 spread between vent and revent in, 270, 271–272
 underground, 241, 246–251
 vertical piping in, 241
cast-iron soil pipe
 cleanout T-branches for, 229, 230
 combination Y and $1/4$ bends for, 221, 222, 225–226
 minimum offsets in, 234, 236, 237–238
 $1/2$ P-traps for, 229, 231
 $1/2$ S-traps for, 229, 231
 $1/4$ bends in, 212, 213, 214, 215
 1/6 bends in, 217, 218
 1/8-bend offsets in, 234–235
 1/8 bends in, 217, 219
 1/16 bends in, 217, 219
 reducers for, 229, 232
 replacing defective section of, 17
 sanitary T-branches for, 217, 220–221
 sweeps for, 213, 216, 217
 T-branches for, 226, 227–228
 upright Y-branches for, 226, 228–229
 vent branches for, 232–233
 Y-branches for, 221, 222, 223–224
Cast Iron Soil Pipe Institute (CISPI), 211
caulking, 158, 161

caulking bead, 96
cellar drains, electric, 48
cement
  applying to plastic pipe, 188–190
  viscosity categories of, 196
center to back, 11
center to center, 11
center to throat, 11
centimeters, converting inches to, 414
chain wrench, 26
channel-lock pliers, 21
check valves
  for discharge piping, 107
  for heater/storage tank connection, 125
  for sewage lift station, 106
  for water heater, 315, 316, 322
chimney vent connector, 319, 320
chlorinated polyvinyl chloride (CPVC) pipe
  applying primer to, 188
  connecting to existing plumbing, 197, 198–200
  cutting, 185–186
  description of, 183
  expansion of, 184
  hot-water heater connections with, 200
  support of, 197
chrome, care and cleaning of, 360
circuit vent, 129
circulating loop, 316
circulating pump
  for heater/storage tank connection, 125, 323–324
  for water heaters, 319, 321, 322, 323–324
city water main, 123
cleaners, for bathtubs, 101
cleanouts
  in diagram of home plumbing system, 212
  for discharge piping, 108
  symbol for, 383
cleanout T-branches, 229, 230
closet augers, 25
closet bend
  bracing for, 248
  in drain-waste-vent (DWV) pipe, 289
closet ell, 290
closet fittings, 288
cold-water headers, 119, 120–122
cold-water inlet, 200
cold-water lines
  for kitchen sinks, 62
  plastic, 184
  recommended length from water heaters, 197
  shock absorbers for, 124, 126–128
  suitable gasket material for, 379
  for water closets, 64–65
  for water heater, 13, 14, 315, 316, 321
combination Y and 1/4 bends
  in cast-iron no-hub systems, 253, 255–257
  in cast-iron soil pipe, 221, 222, 225–226
compression fittings, 208–209
compression stop, 64
compression tank, 5
condensation, on cold-water pipelines, 123
conduction, 7
conduit opening, 316
connections, code requirements for, 155
continuous vent, 129
contraction, allowing for, 203
control stop
  adjusting, 354
  figure of, 75, 349

Index **433**

installing, 345, 346, 347
removing cap from, 349
convection, 7
C.O. plug, 291
copper pipe
  applications for, 279, 282
  for dental offices, 41
  drainage fittings for, 288–297
  fittings and adapters for, 287
  installing in single-family houses, 282–285
  for short-branch connections to fixtures, 283
  Sovent systems for (see Sovent plumbing systems)
  types of, 279, 280–281
  water pressure in, 284
  water pressure losses in, 285–286
corner bathtubs
  electrical connection for, 99
  specifications for, 94
  views of, 92
coupling ring, 116
CPVC pipe
  applying primer to, 188
  connecting to existing plumbing, 197, 198–200
  cutting, 185–186
  description of, 183
  expansion of, 184
  hot-water heater connections with, 200
  support of, 197
crescent wrench, 22
crimped joints, for polybutylene (PB) pipe, 205–207
crimp fittings, 204
crimping tool, 205–207
cross-connections, 123
cross-sectional drawings, of acid-diluting tank installation, 52
cube measurements, conversions for, 417
cylinders
  acetylene, 180
  pressure of oxygen in, 378

D

datum, 3
deaerators
  figure of, 299, 306
  functions of, 297, 300
  parts of, 300
  tied into stack, 300, 301
deaquavators, 33, 35, 36, 37
decimal-fraction equivalents, 398
deep seals, 49
dental chairs
  air piping for, 37
  Den-Tal-Ez Mode CMU-P, 34
  layout for, 36
  supply piping to utility box of, 39
  utility locations for, 41
  vacuum system connections for, 42
dental offices
  air compressor for, 33, 35
  air supply for, 41
  deaquavator for, 33, 35, 36
  dental chair layout for, 36
  drains in, 43
  evacuation system of, 38, 40
  piping in, 33–41
  proper tubing for, 41
  utility stop kit items for, 40
diagrams
  of aerator, 298
  air chambers in, 212
  cleanouts in, 212
  floor drains in, 212
  of flush tank, 64
  of gas-heated water heaters, 13
  of globe valves, 79

434 Index

diagrams *(continued)*
  of impeller, 111
  for installation of burner, 318
  for installation of septic tank, 142
  of lift station, 106
  relief valves in, 212
  riser, 155
  soil pipe in, 212
  water heater in, 212
  See also figures
diaphragm expansion tank, 325, 327
diaphragms, replacing, 81
diaphragm-type flush valve, 73–74
discharge piping, 107
  in sewage pump, 115
  from side of basin, 108
disconnect switch, for water heater, 315, 316
dishwashers, minimum copper tube sizes for short-branch connections to, 283
diverter, 82
double-compartment sinks, 53, 54
downspouts, 17
draft regulator
  for chimney vent connection, 320
  for water heaters, 317, 319, 320
drainage fittings, 288–297
drainage pipe, 379
drainage stacks, 301
drainage systems
  adapters for, 201
  connecting plastic pipe to other materials in, 197, 200–201
  copper pipe for, 282
  plastic pipe used for, 184
  for public building, 131
  purpose of venting in, 130
  for residential building, 132

drainage tee, 303
drain cock, 125
drain fields, 139, 142
  beginning drain lines to, 143
drain lines
  for septic tanks, 142–143
  for water heater, 315, 316
drain pan, 311
drain plug, 61
drains
  building, 10
  in dental offices, 43
  dimensions in corner bathtub, 94
  dimensions in oval bathtubs, 93
  dimensions in whirlpool tubs, 90–91
  electric cellar, 48
  importance of trap to, 63
  pop-up, 61–62
  symbol for, 383
  for whirlpools, 99–100, 102
drain tile
  laying, 143
  for water heater, 315, 316
drain-waste-vent (DWV) pipe, 281
  copper fittings for, 289–295
drain-waste-vent (DWV) system, 211
drawings. *See* working drawings
drawn tube, 279, 280–281
drinking fountains
  approximate water demand for, 285
  example of, 335–336
  features of, 331–333
  minimum copper tube sizes for short-branch connections to, 283
  rough-in of, 336
  symbol for, 383

dry vent, 129
dry well installations, 105
dual vent, 129
dumpy level, 4

E

elastomer bellows, 126
elastomeric sealing sleeve, 238
elbows
  long-turn, 298
  ordering, 8
  in sewer stack, 303
  street, 297
electrical connections, for whirlpools, 98–99
electric cellar drains, 48
electric power tools, grounding, 16
electrolysis, 10
ells, 294
emergency stop, 428
end to center, 11
end to end, 11
engineer's scale, 4, 420–422, 423
escutcheon
  on Aquarian sink fitting, 82
  for single-handle tub and shower set, 88
evacuation system, 38, 40
  connections for, 42
expansion
  in acrylonitrile-butadiene-styrene (ABS) pipe, 203
  allowing for, 203
  in brass pipe, 123
  in chlorinated polyvinyl chloride (CPVC) pipe, 184
  in pipes, 123
  in plastic pipe, 184, 196–197, 202, 203
  in polyvinyl chloride (PVC) pipe, 184
  in steel pipe, 123
  vertical, 202
  of water, 7
expansion fitting, 115

F

face to face, 11
faucets
  approximate water demand for, 285
  installing in kitchen, 84–86
  mounting in service sinks, 85
  no-drip feature of, 80
  repairing, 71–75, 78–80
  on service sinks, 63
  two-handled compression wall-mounted, 83–84
feet
  converting decimal part to inches, 5
  converting inches to decimal part of, 4
  converting to meters, 415
  converting to millimeters, 414
fiberglass outside pump housing, 106, 110
fiberglass-reinforced plastic (FRP) epoxy pipe, 184
fiber washer, 61
figures
  of Act-O-Matic shower head, 366
  of ball cock, 64
  of burner, 13
  of control stop, 75, 349
  of deaerator, 299, 306
  of flush valve, 64
  of lead and oakum joint, 158
  of O-rings, 351, 352
  of refill tube, 64
  of right triangles, 389
  of shower head, 366
  of solar system water heaters, 326–328

436 Index

figures *(continued)*
  of vacuum breaker, 75
  of water cooler, 332
  of water heaters, 312–313
  See also diagrams
filler tube, 64
fire extinguishers, 16–17
fires, fighting, 16
fittings
  compression, 208–209
  for copper pipe, 287
  for drainage, 288–297
  plastic pipe and, 191–197
  polybutylene, 203, 204
fitting socket, 174
fixtures
  approximate heights above floor level for, 9
  connections in bathrooms, 58
  location in bathroom, 57
  symbol for, 383
flanged connection, 107
flanged gate valve, 108
flare fittings, 9, 145
flaring tool, 27
flex tube diaphragm, 351
float, 64, 112
float switches
  for discharge piping, 107
  Mercury, 112
float valve, replacing, 65–66
flood level fin, 60
floor clamp, 250
floor drains
  in diagram of home plumbing system, 212
  for heater/storage tank connection, 125
  symbol for, 383
  venting system for, 136
  water heaters and, 311
floor level, height of fixtures above, 9

flow control orifice, 126
flow resistance, 9
flush connector, 75
flush lever handle, replacing, 66
flushometers
  activating sensor module on, 352
  adjusting control stop on, 354
  adjusting range of, 358–359
  care and cleaning of, 360
  control stop for, 345, 346, 347
  getting too much water with, 365
  handling sensor problems with, 362–364
  installing, 347–348
  installing stop cap on, 354
  operation of, 354–357
  power supply for, 338
  removing components from, 348–350
  replacing batteries in, 359–360
  Sloan G2, 338–342
  testing sensor on, 352, 353, 354
  troubleshooting problems with, 360–361, 362–365
  typical installation of, 345–354
  for urinals, 342–345
flush tanks
  diagram of, 64
  replacing ball cock in, 65–66
  rough-in dimensions for, 343
  water supply for, 65
flush valves
  approximate water demand for, 285
  figure of, 64
  flushometer installation, 345–354
  Optima Plus, 337
  repairing, 337
  RESS model, 337, 338
  Sloan, 70–71, 72, 73–78
  tightening, 65

Index **437**

urinal flushometers, 342
water closet flushometers
(see flushometers)
flux, 165–167
  cleaning off, 170, 171
  for making silver-solder joints, 172
  removing extra, 176
fluxing, 174, 175
flux pocket, 177
folding rule, 380, 382
formulas
  for bend angles of pipe, 390–392
  for determining area, 394
  for determining capacities, 393
  for determining pressures, 394–395
  for finding diagonal of square, 388–389
  Pythagorean Theorem, 389
  for right triangles, 389–393
  for rolling offsets, 395–398, 406–407
  for square roots, 385–388
  for temperature conversions, 416
fraction-decimal equivalents, 398
fresh-air systems, working drawings for, 45–46
fresh-air vents, 130
friction clamp, 250
friction welding, 209

G

garbage disposals
  branch waste line for, 130, 132
  emptying of, 130
  flushing of, 55
  parts used for mounting, 54
  proper use of, 55
  residential, 49–50, 53–55
gas burner
  figure of, 13
  lighting, 320–322
gas-charging device, 126
gases, found in sewer air, 10
gas-heated water heaters, 314–315, 317
  diagram of, 13
  lighting, 14
gaskets
  on gas burner, 318
  suitable material for, 379
gasoline conduction, 379
gas piping
  suitable gasket material for, 379
  for water heaters, 320
gate valves, 9
  for discharge piping, 107
  for heater/storage tank connection, 125
  for pressure-reducing valve, 124
globe valves, 9
  diagram of, 79
glycerine hydraulic displacement fluid, 126
grade levels, shooting, 4–5
grams, measurement equivalents to, 416
grease pipe, 110
grease traps, 45, 47
ground fault circuit interrupter (GFCI), 98
G2 flushometer, 338–342
  electrical specifications for, 344
  installing flush volume regulator for, 350–351
  operation of, 356
  range adjustment for, 358–359
  rough-in for, 344
  for urinals, 342–345
  water pressure in, 345
guide bearing, 115

H

hangers
  determining location of, 119
  for horizontal piping, 241, 243–245
  sizes of, 382, 384
Hastings Drinking Fountain, 335–336
headers, 119, 120–121
head pressure, 423
head stub, 151, 152
heat
  required to change ice to liquid water, 8
  transfer of, 7
heater, connection to storage tank, 124, 125
heat exchangers, 330
heat-fusion joints, 207–208
heating systems
  compression tank for, 5
  copper pipe for, 282
  expansion of water in, 7
high-density insulation, 325
high-density magnesium anodes, 325
high-water alarm, 112
hinged access panel, for water heater, 316
hoisting signals, 427–428
horizontal joints
  making, 170–171
  soft soldering, 177, 179
horizontal piping
  in cast-iron no-hub system, 241, 243–245
  support for, 241, 243–245
  underground, 241, 246–251
horizontal round joints, 145–148
horizontal runs, 195
horizontal stack offset, 299, 301
horizontal stringers, 16
hot-water headers, 119, 120–122
hot-water heater connections, 200
hot-water lines
  heat loss from, 7
  for kitchen sinks, 62
  plastic pipe for, 184
  recommended length from water heaters, 197
  shock absorbers for, 124, 126–128
  suitable gasket material for, 379
hot-water outlet, 321
hot-water supply, 332–333. *See also* hot-water lines; water heaters
Hustler II compressor, 33, 35
  air piping for, 37
hydronic heating, 282
hygrometer, 17

I

impeller
  reference diagram for, 111
  for sewage lift station, 106
  in sewage pump, 115
inches
  converting decimal parts of foot to, 5
  converting to centimeters, 414
  converting to decimal parts of foot, 4
  converting to meters, 415
  converting to millimeters, 412, 413
increaser-reducer, 273, 275
individual vent, 129
in-line offset, 297, 301
internal wrench, 23
invert elevation, 3–4
iron, melting point of, 379

J

job brass ferrule, 151, 152
joints

applying solder to, 148, 149–150, 177
assembling in plastic pipe, 190–191
bell-and-spigot, 160
for bell-and-spigot soil pipe, 161
butt, 153
code requirements for, 155
crimped, 205–207
heat-fusion, 207–208
horizontal, 170–171
horizontal round, 145–148
lap, 152–153
lead and oakum, 157–162
in lead pipe, 161, 162
between plastic and metals, 196–197
in plastic tubing, 185–191
for polybutylene (PB) pipe, 205–207
in polyvinyl chloride (PVC) pipe, 187
preparation for making, 172, 173–176
in pump suspension system, 114
putting in bell end pipe, 185–190
set time of, 191, 192
silver-solder, 172, 173–176
solvent weld, 196–197
test-fitting, 186–187
vertical, 170–171
joules, conversion to British thermal units (Btu), 417, 418

K

Kem-Temp polyvinylidene fluoride (PVDF) pipe, 184
kilograms (kg), measurement equivalents to, 416
kilopascals (kPa), 376–377
kitchen sinks
installing faucets in, 84–86
minimum copper tube sizes for short-branch connections to, 283
symbol for, 383
waste line for, 62
water lines for, 62
knots, 425–426

L

ladders, proper use of, 16
lap joints, 152–153
laundry tray
minimum copper tube sizes for short-branch connections to, 283
symbol for, 383
lavatories
approximate water demand for, 285
headers for, 119, 120–121
installing pop-up drain in, 61–62
minimum copper tube sizes for short-branch connections to, 283
rough-in of, 59, 60
sanitary, 61
symbol for, 383
venting system for, 135, 137
vents for, 133, 212
lavatory vent, 212
lead, melting point of, 379
lead and oakum joints
and connections, 161–162
figure of, 158
information on, 159
preparing, 157–159
lead-caulking, 200
lead joints, 161
lead pans, 66
lead pipes
applying solder to joints of, 148, 149–150
beveling end of, 146

440 Index

lead pipes *(continued)*
  butt joints on, 153
  cleaning ends of, 147
  fitting together, 147–148
  flaring ends of, 145
  joints in, 162
  lap joints on, 152–153
  lead-joining work on, 151–152
  preparing horizontal round joints for, 145–148
  preventing oxidation on, 153
  stack venting of, 154–155
  wiping head stub on job brass ferrule, 151, 152
  and wiping solder, 145
leaks, checking bathtubs for, 92
leveling rod, 4–5, 6
lift stations
  adjusting impeller in pump for, 110–111, 114
  average sewage flow rate in, 118
  construction features of pumps for, 114–115
  diagram of, 106
  discharge piping layout for, 107
  fiberglass pump housing for, 106, 110
  flow quantities in, 118
  Mercury float switch for, 112
  outside piping of, 105–110, 111, 112, 113
  outside pump housing for, 113
  piping discharging from side of basin, 108
  types of, 105
  and wastewater collection systems, 115, 118
lift wires, 64
lime, rinsing from eyes, 16
line temp control, 316, 322
liquid volume, 411
liters, kilogram equivalent, 416
local vent, 129
locking ring, tightening, 353
long-turn 90° elbow, 298
loop vent, 139–140
  defined, 129–130
louver
  for pump housing, 113
  for sewage lift station, 106
low consumption urinals, 340
lower load, 428
low-hub 1/4 bends, 213, 215
lube pipe, 115

M

magnesium anodes, 325
main shutoff valve, 128
main vent, 129
manhole, for discharge piping, 107, 108
manifold, 82
measurements
  metric abbreviations for, 410
  of pipes, 8–9
  See also metric system
melting points, 379
mercury, pressures of, 377
Mercury float switch, 106, 112
meters, converting inches and feet to, 415
metric scale, 420–422, 423
metric system, 376–377
  abbreviations for, 410
  acceptance of, 409
  conversions to, 412–416, 417
  liquid volume in, 411
  measuring length in, 411
milliliters (mL), gram equivalent to, 416
millimeters (mm)
  converting feet to, 410
  converting inches to, 412, 413
mixing chamber, 297
mixing valve, 315, 316, 322

mobile homes, working drawing of connection for, 49, 51
mountings, for residential garbage disposal, 54
multistory stacks, installing, 201–202

N

National Sanitation Foundation, 196
NIBCO, CPVC-to-metal unions and adapters, 198–199
nipples, 13, 14
nominal pipe size (N.P.S.), 8
nosepiece, 298

O

oakum, packing into bell-and-spigot, 160
offset closet fittings, 288
offsets
 double in-line, 297, 301
 horizontal stack, 299, 301
 1/8 bend, 234–235
 using 1/6-, 1/8-, and 1/16-bend fittings, 234, 236, 237–238
oil lines, suitable gasket material for, 379
$1/2$ P-traps, 229, 231, 273, 274
$1/2$ S-traps, 229, 231, 273, 274
$1/4$ bends
 in cast-iron no-hub systems, 261–262
 in cast-iron soil pipe, 212, 213, 214, 215
 combined with Y bends, 221, 222, 225–226, 253, 255–257
 double, 266, 269–270
 short-radius, tapped, 266, 270, 271
 using with wye, 270, 271–272

1/5 bends
 in cast-iron no-hub systems, 258
 minimum offsets using, 276
1/6 bends
 in cast-iron no-hub systems, 258
 in cast-iron soil pipe, 217, 218
 minimum offsets using, 277
 using formulas to determine, 392
1/8-bend offsets, 234–235
1/8 bends
 in cast-iron no-hub systems, 258, 259
 in cast-iron soil pipe, 217, 219
 minimum offsets using, 277
1/16 bends
 in cast-iron no-hub systems, 258, 259
 in cast-iron soil pipe, 217, 219
 minimum offsets using, 277
Optima Plus flushometer
 activating sensor module on, 352
 adjusting control stop on, 354
 adjusting range of, 358–359
 assembling flex tube diaphragm in, 351–352
 8100 series, 337, 339
 testing sensor on, 352, 353, 354
O-rings
 on Aquarian sink fitting, 82
 figure of, 351, 352
 identifying wear and tear in, 79
 in Optima Plus flush valve, 349
ounces, measurement equivalents to, 416
oval bathtubs
 specifications for, 93
 views of, 92
overflow
 dimensions in corner bathtubs, 94
 dimensions in oval bathtubs, 93
 dimensions in whirlpool tub, 90–91

overflow *(continued)*
  installing in bathtub, 67–69
  in water heater, 64
  for whirlpools, 102
overflow strainer, 108
oxyacetylene welding
  safety measures for, 179–180
  temperature of flame for, 378
oxygen systems, 282

P

packing nut, 79
paint, removing, 101
partial vacuum, 9
parts lists
  for Aquaseal sink fitting, 77–78
  for Sloan flush valve, 73–76
PB pipe, 203–205
  applications of, 205
  crimped joints for, 205–207
  description of, 183
PE pipe, 183
phase convertibility, 311
pipe compound, 88
pipe continuity, 163
pipe cutters, 20, 185–186
pipe dopes, 195
pipelines
  air chambers in, 123
  condensation on, 123
  corporation stop on, 123
  cross-connections in, 123
  hung from ceiling, 119–124
  invert of, 123
  pressure regulator for, 124
  shock absorbers for, 124, 126–128
  slanting, 123
  water hammer in, 123
pipe plug
  on Aquarian sink fitting, 82
  function of, 163

pipes
  calculating proper angles for, 390–393
  changing direction of, 163
  cleaning fitting socket of, 174
  connecting plastic with metal, 196–197, 198–199
  in dental offices, 33–41
  expansion of (*see* expansion)
  forming angle in, 380, 382
  hoisting of, 427–428
  lead (*see* lead pipes)
  making knots for lifting, 425–426
  nominal size of, 163
  preparing lead and oakum joints in, 157–159
  proper hanger sizes for, 382, 384
  removing burrs from, 173
  replacing defective section of, 17
  risers, 119
  running (*see* pipelines)
  size measurement of, 8–9
  stacks, 119
  standard length of, 163
  threading machine for, 22
  thread measurements for, 380
  water capacity of, 123
  for water supply, 8
pipe wrench, 19–20
piping systems, regulating flow in, 163–164
plaster, removing, 101
plastic pipe
  allowing for expansion and contraction in, 203
  applications for, 184
  applying cement to, 188–190
  applying primer to, 188
  approval of, 183
  assembling, 17
  assembly of joint in, 190–191

Index **443**

compression fittings for, 208–209
connecting with metal pipe, 196–197, 198–199
cutting, 185–186
expansion in, 184, 196–197, 202, 203
and fittings, 191–197
heat-fusion joints in, 207–208
joints with, 185–191
lead-caulking of, 200
for multistory stacks, 201–202
National Sanitation Foundation's standards for, 196
and pipe dopes, 195
polybutylene, 203–205
pressure tests for, 191–195
preventing vertical expansion in, 202
removing gloss from, 187
smoothing end of, 186
storage of, 194–195
supporting systems of, 202
test-fitting joint in, 186–187
threading of, 201
tubing cutter for, 206
types of, 183–184
water pressure ratings for, 193–194
for whirlpools, 100
wiping excess cement off of, 192
pliers, 21
plug cock
between storage tank and water heater, 323–324
for water heater, 316, 322
plugs
on Aquarian sink fitting, 82
function of, 163
plumb bob, 119
plumbers
abbreviations used by, 375
basic tools of (*see* tools)
common terms to remember, 11
hoisting signals for, 427–428
knots used by, 425–426
safety precautions for, 15–17
plumbers' putty, 85
for drain installation, 61
plumbers' tape, 85
plumbing
defined, 15
tips for, 17
plumbing systems
polybutylene, 203–205
single-stack (*see* Sovent plumbing systems)
use of cast-iron pipe in, 212
plumbing tools. *See* tools
plumbing tube, 279
polybutylene (PB) pipe, 203–205
applications of, 205
crimped joints for, 205–207
description of, 183
polyethylene (PE) pipe, 183
polypropylene pipe, 184
polyvinyl chloride (PVC) pipe
allowing for expansion and contraction in, 203
applying primer to, 188
assembling, 17
connecting to other materials, 197, 200–201
cutting, 185–186
description of, 183
drawing of joint in, 187
expansion of, 184
for multistory stacks, 201–202
removing gloss from, 187
solvent cements for, 196
storage of, 194–195
water pressure ratings for, 193
for whirlpools, 100
pop-up drain, 61–62
pounds (lb), measurement equivalents to, 416

pounds per square inch (psi)
  kilopascal equivalent, 419
  water equivalents to, 376
power tools, grounding, 16
pressure
  absolute, 7
  for columns of water, 419, 423
  conversion chart for, 419
pressure gage, for pressure-reducing valve, 124
pressure loss, 285–286
pressure-reducing valves
  function of, 123
  installation of, 124
pressure regulator, 124
pressure-relief line
  in Sovent system, 299
  tied below horizontal offset, 301
  for waste stack, 304
pressure-relief outlet, 298
pressure-relief system, 299
pressure-relief valve
  for boiler, 125
  for storage tank, 125, 323–324
  for water heater, 13, 316, 321
pressurized argon pneumatic displacement cushion, 126
primary treatment, 10
primer, 188
priming, of plastic pipe, 185, 188
propylene glycol, 325
protector rod, 8
P-traps
  in bathtub, 66
  for cast-iron no-hub system, 273, 274
  for drains in dental offices, 43
  for mobile homes, 49, 51
  $1/2$, 229, 231
P-trap seal, 47, 49, 50

pump casing, 115
pumps
  circulating, 319, 321, 322, 323–324
  construction features of, 114–115
  housing for, 113
  for lift stations, 105, 109, 110–111, 113, 114
  for sewage and wastewater applications, 116–117
  submersible, 116–117, 118
  vertical, 110–111, 114
pump shaft, 114
PVC pipe
  allowing for expansion and contraction in, 203
  applying primer to, 188
  assembling, 17
  connecting to other materials, 197, 200–201
  cutting, 185–186
  description of, 183
  drawing of joint in, 187
  expansion of, 184
  for multistory stacks, 201–202
  removing gloss from, 187
  solvent cements for, 196
  storage of, 194–195
  water pressure ratings for, 193
  for whirlpools, 100
Pythagorean Theorem, 389

Q

Qicktite compression fitting, 208–209

R

radiation, 7
rainwater leaders, 17
rectangular whirlpool tubs
  specifications for, 90–91
  views of, 89

reducers
    for cast-iron soil pipe, 229, 232
    for discharge piping, 108
reducing elbows, 8
refill tubes
    figure of, 64
    installing, 66
reflow cutoff valve, 106
relative humidity, measuring, 17
relief line
    using chlorinated polyvinyl chloride (CPVC) pipe for, 197
    for water heater, 13
relief valves
    in diagram of home plumbing system, 212
    for gas-heated water heaters, 13, 14
    placement of, 10
    relief line from, 10
    temperature-sensitive element of, 10
    temperature setting of, 314
    vacuum, 9
    for water heaters, 13, 200, 311, 314
relief vent, 129
repair kits
    for Aquaseal valve, 81
    for flush valves, 73
RESS model flush valves, 337, 338
    installing flush volume regulator for, 350–351
    installing stop cap on, 354
    retrofit installations for, 348–350
restraining fittings, 202
retainer, 82
revent, 270, 271–272
right triangles
    figure of, 389
    working with, 389–393

riser diagrams, 155
risers, 119
rolling offsets
    calculating, 395–397
    constants for calculating, 407
    simplified calculation of, 397–398, 406–407
    table of measurements for, 399–405
roof drain, symbol for, 383
roof flashing, 212
rough-ins
    for bathroom fixtures, 57
    for bathtubs, 66, 67
    for connections for mobile home, 51
    for drinking fountains, 336
    for flushometers, 339, 342
    for flush tanks, 343
    for garbage disposal installation, 53
    for G2 flushometer, 344
    for lavatory, 59, 60
    for sanitary tee, 53
    for service sink, 85
    for washing machine, 83
    for water closet, 59, 64
    for water coolers, 333–334
    for whirlpool tubs, 95

S

safety precautions, 15–17
sand traps
    places that use, 46
    working drawing of, 46–47, 49
sanitary drains, 63
sanitary lavatories, 61
sanitary T-branches
    for cast-iron no-hub systems, 262, 263–264
    for cast-iron soil pipe, 217, 220–221
    tapped, 263, 265–266

446 Index

sanitary tees
  rough-in of, 53
  in sewer stack, 303
scald prevention, 323–324
scale rules, 12
screwed unions, 163
screws, reference information for, 379–380, 381
seat-dressing tool, 78–79
secondary treatment, 10
sectional level, 3
sensor module
  activating, 352
  testing, 352, 353, 354
septic tanks
  bacterial action of, 132, 135, 142
  cleanout of, 135, 138
  drain fields for, 139, 142
  drain lines for, 142–143
  installation diagram of, 142
  location of, 142
service access, for whirlpool bathtubs, 96–98
service sinks
  minimum copper tube sizes for short-branch connections to, 283
  mounting faucet in, 85
  symbol for, 383
  two-handle faucet for, 84
  waste line for, 63
  water lines for, 63
set times, 191, 192
sewage
  lift stations for (see lift stations)
  pumping stations for, 105
sewage-ejector vents, 130
sewer gases, interceptors for, 45
sewers
  building, 10
  gases found in, 10
  invert of, 123
  schematic drawing of system of, 58
sewer tie-ins, 303–304
shave hooks, 151, 152–153
shock absorbers, 123, 124, 126–128
shower heads
  adjusting direction of, 368
  features of, 361
  figure of, 366
  installation of, 366–368
  measurements for, 370
  minimum copper tube sizes for short-branch connections to, 283
  parts list for, 369
  pre-installation considerations for, 361
  repair kit for, 371
  water lines for, 361
shower stalls
  lead pans for, 66
  symbol for, 383
shutoff valve, 64
  for water heater, 315, 316
SI. *See* metric system
silver brazing
  expense of, 164
  feeding alloy for, 167, 169
  fluxing during, 165–167
  heating tube for, 166–169
  of horizontal and vertical joints, 170–171
  preliminary steps for making joints by, 172, 173–176
  proper torch heat for, 164–170
  purpose of, 164
  stages of flux in, 165
single-handle tub and shower set, 87, 88
single-stack plumbing systems. *See* Sovent plumbing systems

sinks
  Americans with Disabilities Act (ADA) guidelines for, 62–63
  approximate water demand for, 285
  Aquaseal fitting for, 77–78
  in kitchen, 62–63
  mounted, 63
  placement specifications for, 60
  service, 63
  symbol for, 383
  two-handle compression wall-mounted faucets for, 83–84
siphon-jet urinals, 340
slab-on-grade, 249
slanting pipelines, 123
sleeve coupling, 239
sleeve guide bearing, 110
Sloan Valve Company
  Act-O-Matic shower head, 361, 366–371
  adjustable tailpiece from, 347–348
  control stop from, 345, 346, 347
  flushometers for urinals from, 342–345
  flush valves, 70–71, 73–78
  G2 flushometer, 338–345
  repairing flush valves from, 337
soft soldering
  guidelines for, 172, 177, 179
  purpose of, 164
soil branches
  connected below deaerator fitting, 305–308, 309
  connected into pressure-relief line, 304–305, 309
  in sewer stack, 303
soil pipe, 155
  cast-iron (see cast-iron soil pipe)
  in diagram of home plumbing system, 212
solar system water heaters
  Conservationist, 325–329
  figures of, 326–328
  heater exchanger for, 330
  specifications for, 329
solder
  applying, 178
  applying as joint cools, 177
  applying to joints, 148, 149–150
  for butt joints, 153
  and flux pockets, 177
  holding back water while applying, 17
  melting point of, 379
  preventing splashing of, 147
  wiping, 145
soldering irons
  for butt joints, 153
  for lead-joining work, 151–152
solvent cement, 188–189
  viscosity categories of, 196
solvent weld joints, 196–197
Sovent plumbing systems
  aerator fittings for, 296–299
  cost-saving potential of, 308–309
  deaerator fittings for, 297, 300
  first installation of, 286
  fittings for, 308
  function of, 286
  invention of, 286
  parts of, 296
  picture of installation of, 307
  schematic drawing of, 302
  sewer and waste tie-ins in, 303–304
  stack anchoring in, 305
  stacks in, 300, 308
spiral ratchet pipe reamer, 25
spout, 82
spray head, 82
spray hose, installing, 86

448 Index

spud coupling, 75
spud flange, 75
spud wrench, 24
square knot, 426
square measurements, conversions for, 417
square roots, finding, 385–388
squares
  of common numbers, 387
  finding diagonal of, 388–389
stack inlet, 297, 298
stack offset, 299, 301
stack outlet, 297, 298
stacks
  anchoring, 305, 309
  defined, 119
  design features of, 301
  horizontal, 304
  installing multistory, 201–202
  schematic drawing of, 302
  sewer and waste tie-ins on, 303–304
  soil and waste connections in, 305
  tying together, 300, 308
  vertical, 304
stall urinals, 69–70
steel, melting point of, 379
steel pipe, expansion of, 123
stop coupling, 75
stopper, 61
stop signal, 428
stop valve, 200
storage tanks, 8
  connection to heater, 124, 125, 323–324
straight nips, 28
straight pipe wrench, 20
strainer, 106
S-traps
  for cast-iron no-hub system, 273, 274
  $1/2$, 229, 231, 273, 274

strap wrench, 24
street elbows, 297
submersible pump, 116–117, 118
suction cover, 102
suction plate
  for impeller, 111
  for sewage lift station, 106
sump pump, 212
supply flange, 75
supply pipes
  shock absorbers for, 124, 126–128
  of water (*see* water supply lines)
  for water closet flush valves, 64
swaging tools, 21
sway brace, 251
sweat solder applications, 345, 346, 347
sweeps
  in cast-iron no-hub systems, 259–261
  in cast-iron soil pipe, 213, 216, 217

T

tailpiece, 75
tank ball, 64
tapped extension place, 273, 274–275
tar, removing, 101
T-branches
  for cast-iron no-hub systems, 262, 263–264
  for cast-iron soil pipe, 226, 227–228
  cleanout, 229, 230
  sanitary, 217, 220–221, 262, 263–266
tees
  for drain-waste-vent (DWV) pipe, 293
  ordering, 8

in sewer stack, 303
test, 261, 262, 263
Teflon tape, 88
temperature relief valve
  for storage tank, 125, 323–324
  for water heater, 13, 316, 321
temperature systems, conversions for, 416–417, 418
tempered water outlet, 315, 316, 322
test-fitting, 186–187
test tees
  for cast-iron no-hub systems, 261, 262, 263
  with plug, 295
thermometer, for water heater, 315, 316
thermostat, for storage tank, 125
threaded plug, 126
threading machine, 22
timber hitch knot, 426
tin, melting point of, 379
tongue-and-groove alignment, 110
tools
  adjustable wrench, 22, 85
  ballpeen hammer, 19
  basin wrench, 23, 85
  chain wrench, 26
  channel-lock pliers, 21
  closet auger, 25
  for crimped joints, 205–207
  flaring tool, 27
  folding rule, 380, 382
  for heat-fusion joints, 207–208
  for installing kitchen faucet, 85
  for installing single-handle tub and shower set, 87
  internal wrench, 23
  lever-type tube benders, 27
  for making lead and oakum joints, 157
  pipe and bolt threading machine, 22
  pipe cutters, 20, 185–186
  pipe wrench, 19–20, 85
  plumb bob, 119
  plumbers' putty, 85
  plumbers' tape, 85
  shave hooks, 151, 152–153
  spiral ratchet pipe reamer, 25
  spud, 24
  straight nips, 28
  straight pipe wrench, 20
  strap, 24
  swaging tools, 21
  torque wrench, 26
  tubing cutter, 24
torches
  safety measures for oxyacetylene welding, 179–180
  for silver brazing, 164–170
  for soft soldering, 172, 177, 179
  temperature of flame of, 378
torque wrench, 26
trailers, working drawing of connection for, 49, 51
transfer of heat, 7
transition unions, 197, 198–199, 200
trap arm, 303
traps, 212
trap seals
  protection from evaporation, 47
  protection of, 196
  working drawings for, 47, 49, 50
travel backward, 428
travel forward, 428
trip lever, 64
trip waste, 67–69
tub and shower set, installing, 87, 88
tube benders, 27
tube tails, 61
tubing cutters, 24
  for plastic pipe, 206
tub spouts, 88

450 Index

twin ells, 294
two-handle service sink faucet, 84
type K copper pipe, 280
type L copper pipe, 280
type M copper pipe, 281

U

U-frame skirt mounting detail, 96
unions
   CPVC-to-metal, 198–199
   function of, 163
   for pressure-reducing valve, 124
   for storage tank, 125
   for water heater, 315, 316
upright Y-branches, 226, 228–229, 266, 268–269, 292
urinals
   checking operation of flushometer on, 355
   flushometers for, 340, 342–345
   minimum copper tube sizes for short-branch connections to, 283
   operation of flushometer on, 357
   stall, 69–70
   symbol for, 383
   venting system for, 136
utility layout, 40
U-trap, 63

V

vacuum, 9
vacuum breaker, 9
   coupling for, 349
   figure of, 75
   installing flush connection for, 347, 348
vacuum relief valve, 9
valve controls, locating, 127–128
valves
   Aquaseal, 80, 81
   check (*see* check valves)
   diaphragm-type flush, 73–74
   flanged gate, 108
   float, 65–66
   flush (*see* flush valves)
   gate, 9, 107, 124, 125
   globe, 9, 79
   locating controls for, 127–128
   main shutoff, 128
   mixing, 315, 316, 322
   no-drip, 80
   pressure-reducing, 123, 124
   pressure-relief, 13, 125, 316, 321, 323–324
   reflow cutoff, 106
   relief (*see* relief valves)
   repairing, 71–72, 78–80
   shutoff, 64, 315, 316
   for single-handle tub and shower set, 87
   stop, 200
   temperature relief, 13, 126, 316, 321, 323–324
   tightening in water closet, 65
   vacuum relief, 9
vent branches, 232–233
vent connector, 319, 320
vent cross, 294
venting systems
   abbreviations used in drawings of, 132
   for bathtubs, 137, 141
   copper pipe for, 282
   examples of, 130
   for floor drains, 136
   for lavatories, 133, 135, 137
   looped, 139–140
   for public building, 131
   purpose of, 130
   for residential building, 132
   schematic drawing of, 58
   for urinals, 136
   using waste and vent stacks, 138

Index **451**

for wall-hung water closets, 134
for water closets, 133, 135, 136, 137
for water heaters, 318–319, 320
vent roof flashing, 162
vents
back, 129
branch, 129, 232–233
in chimney, 319, 320
circuit, 129
connector for, 319, 320
continuous, 129
for discharge piping, 107
dry, 129
dual, 129
fresh-air, 130
individual, 129
for lavatories, 133, 212
local, 129
loop, 129–130, 139–140
main, 129
relief, 129
revent, 270, 271–272
roof flashing, 162
sewage-ejector, 130
Sovent system of (see Sovent plumbing systems)
spread between revent and, 270, 271–272
stack of, 129, 138
types of, 129–130
for water closets, 133
for water heater, 13
wet, 129
yoke, 129
vent stack, 129, 138
vertical braces, 16
vertical joints, 170–171
vertical piping, 241, 242
vertical pumps
adjusting impeller for, 110–111, 114

cleaning basin for, 110
general instructions for, 110–111, 114

W

wall-hung water closets, 64
wash-down urinals, 340
washing machines
minimum copper tube sizes for short-branch connections to, 283
rough-in of, 83
waste branches
connected below deaerator fitting, 305–308, 309
connected into pressure-relief line, 304–305, 309
in sewer stack, 303
waste inlet, 297, 303
waste line
for garbage disposal, 130, 132
for kitchen sinks, 62
for service sinks, 63
for water cooler, 333
waste stack, 138
termination of, 130
waste tie-ins, 303–304
wastewater collection systems, 115, 118
water
air pressure needed to elevate, 376–377
boiling points of, 378
calculating demand for, 284–285
in drain-waste-vent (DWV) system, 211
expansion in heating system, 7
heating of (see water heaters)
heat required to change from ice to liquid, 8
kilopascals (kPa) of, 376
measurement of, 376–377

water *(continued)*
  pounds per square inch (psi) of, 376
  pressure needed to elevate, 420
  for stall urinal, 69
  supply of *(see* water supply lines)
water closets
  auger for, 25
  checking operation of flushometer on, 355
  diagram of flush tank for, 64
  flushometers for, 338–342
  flushometers for high rough-in installations of, 339, 342
  installing to closet floor flange, 65
  in looped vent system, 139
  minimum copper tube sizes for short-branch connections to, 283
  operation of flushometer on, 356
  replacing float valve in, 65–66
  replacing flush lever handle on, 66
  rough-in of, 59, 64
  symbol for, 383
  tightening valves for, 65
  venting system for, 134, 135, 136, 137
  vents for, 133
  wall-hung, 64
  water lines for, 64–65
water coolers
  features of, 331–333
  figure of, 332
  mounting, 334
  rough-in of, 333–334
  wall-hung, 332
water distribution systems
  pressure losses in, 285–286
  using copper pipe for, 282–185
water faucets. *See* faucets
water hammer arrestors, 123

water hammer
  reducing, 123
  suppressing, 124, 126–128
water heaters
  BTP model of, 313
  component failure in, 314
  connected to storage tank, 323–324
  connecting CPCV pipe to, 197, 200
  connecting water lines to, 319, 321, 322, 323–324
  corrosion in, 314
  in diagram of home plumbing system, 212
  distance from unprotected wood, 8
  drain pan for, 311
  DVE and DRE models of, 312
  figures of, 312–313
  gas, 13, 14, 314–315
  gas piping for, 320
  heater outlet for, 314
  heater water inlet for, 314
  installation of burner in, 318
  lighting burner for, 14, 320–322
  maximum acceptable temperature for, 7
  one-temperature, 316
  preventing corrosion in, 8
  proper location of, 311, 314
  proper ventilation for, 318–319, 320
  relief valves for, 314
  solar systems, 325–330
  two-temperature, 315
water inlet, 315, 316
water level, 103
water lines
  connecting to water heaters, 319, 321, 322, 323–324
  flushing, 88
  for kitchen sinks, 62
  for service sinks, 63

Index **453**

for shower heads, 361
for water closets, 64–65
water mains, 282
water outlet, for water heater, 315, 316
water pressure
 calculating, 394–395
 and copper pipes, 284
 in dental offices, 43
 in flushometers, 338, 345
 friction causing loss of, 285–286
 measurement of, 419
 ratings for plastic pipe, 193–194
 relief system for, 297, 299
 for shower heads, 361
water service meter, 212
water softener, 8
water supply lines
 adequate pressure in dental offices, 43
 connecting, 88
 cutting for control stops, 345
 flushing, 347
 for flush tanks, 65
 for lavatory, 60
 pipes for, 8
 risers, 119
 for shower head, 368
 shut off for, 86, 88
 sliding covering tube onto, 347
 sliding threaded adapter onto, 345
 underground pipe for, 279
 for water cooler, 333
 water pressure in, 211
 for whirlpools, 100
 See also cold-water lines; hot-water lines
weight, measurement conversions for, 416
welding
 friction, 209
 oxyacetylene, 179–180, 378
 solvent, 196–197

wet vent, 129
wet-well installation, 105
whirlpool bathtubs
 access to service connections, 96–98
 air induction controls for, 104
 checking for leaks in, 92
 controlling action of, 103–104
 corner, 92, 94
 drains for, 99–100
 electrical connections for, 98–99
 fittings for, 102
 inspecting shell of, 89, 92
 on-off switch for, 102, 103
 operation of, 101–102
 oval, 92, 93
 post-installation cleanup of, 101
 pre-installation considerations for, 87, 89, 92
 rectangular, 89, 90–91
 rough-ins of, 95
 U-frame skirt mounting detail for, 96
 water jets in, 102, 104
 water level in, 103
 water supply for, 100
wiping solder
 cleaning, 148, 151
 proper heat for, 148
 working with, 145
working drawings
 for acid-diluting tanks, 52
 for electric cellar drains, 48
 for fresh-air systems, 45–46
 for residential garbage disposals, 49–50, 53–55
 for sand traps, 46–47, 48
 for trailer connections, 49, 51
 for trap seals, 47, 49, 50
wrenches
 adjustable, 22, 85
 basin, 23, 85
 chain, 26
 internal, 23

454 Index

wrenches *(continued)*
  pipe, 19–20, 85
  spud, 24
  straight pipe, 20
  strap, 24
  torque, 26
Wrightway heat-fusion tool, 207, 208
wyes, 270, 271–272

Y

Y bends
  combined with $1/4$ bends, 221, 222, 225–226, 253, 255–257
  using formulas to determine, 392

Y-branches
  in cast-iron no-hub systems, 253, 254, 255
  in cast-iron soil pipe, 221, 222, 223–224
  with C.O. plug, 291
  in drain-waste-vent (DWV) pipe, 293
  tapped, 266, 267–268
  upright, 226, 228–229, 266, 268–269, 292
yoke vent, 129

Z

zinc, melting point of, 379

Made in the USA
Coppell, TX
22 May 2024